Fine Tuning Air Conditioning Systems and Heat Pumps

Lee A. Miles

Business News Publishing Company
Troy • Michigan

Copyright © 1995
Business News Publishing Company

All rights reserved. Except as permitted under the United States Copyright Act of 1976, no part of this publication may be reproduced or distributed in any form or means, or stored in a database or retrieval system, without the prior written permission of the publisher, Business News Publishing Company.

Library of Congress Cataloging in Publication Data

Miles, Lee A.
    Fine tuning air conditioning systems and heat pumps / Lee A. Miles.
      p. cm.
    ISBN 0-912524-96-0
    1. Air conditioning--Equipment and supplies--Maintenance and repair. 2. Heat pumps--Maintenance and repair. I. Title.
TH7687.7.M55   1995   94-30563
697.9'3'0288--dc20   CIP

Editor: Joanna Turpin
Art Director: Mark Leibold
Copy Editor: Carolyn Thompson

This book was written as a general guide. The author and publisher have neither liability nor can they be responsible to any person or entity for any misunderstanding, misuse, or misapplication that would cause loss or damage of any kind, including loss of rights, material, or personal injury, or alleged to be caused directly or indirectly by the information contained in this book.

Printed in the United States of America
7 6 5 4 3 2 1

TABLE OF CONTENTS

CHAPTER ONE
Thermal Balance and Psychrometrics 1

CHAPTER TWO
Adjusting Airflow 11

CHAPTER THREE
Determining Refrigerant Charge 31

CHAPTER FOUR
Determining System Capacity 43

CHAPTER FIVE
Adjusting for Proper Local Performance 57

CHAPTER SIX
Troubleshooting 65

Glossary of Technical Terms 75

Abbreviations 87

Answers 91

CHAPTER ONE
Thermal Balance and Psychrometrics

A refrigeration system transfers heat from one place, a *heat source*, to another place, a *heat sink*. The heat source is any material — air, liquid, or solid — from which it is necessary to remove heat to control temperature, humidity, or a combination of the two. Systems designed strictly for refrigeration, such as medium-temperature grocery display cases and walk-in coolers, usually control only temperature; whereas air conditioning systems and heat pumps control both temperature and humidity.

In air conditioning systems and heat pumps, temperature is controlled automatically and humidity is controlled by adjusting the system manually. But in some cases, humidistats are incorporated into their control systems for more control over humidity levels. A *humidistat* is a humidity-sensing control that turns the humidifier on and off in cycles. A *humidifier* adds water vapor to the air.

In order to produce the required levels of temperature and humidity for maximum comfort conditions, system adjustments are necessary. The objective of this text is to detail these various adjustments and explain how and why they are performed.

To understand the actions and reactions that take place in refrigeration systems used in air conditioning systems and heat pumps, an understanding of thermal balance is required.

THERMAL BALANCE

Heat is absorbed from the heat source by the *evaporator coil*, otherwise known as the direct expansion (DX) coil, by boiling or evaporating a liquid refrigerant in the coil. The heat is transferred to the *condenser*, where compressed refrigerant vapor is cooled until it becomes a liquid. The liquid is disposed of in the heat sink.

During compression, electrical energy is used to produce the temperature at which the refrigerant vapor condenses. Therefore, the total heat that enters the heat sink is made up of the heat removed by the refrigeration action plus the heat energy equivalent of the electrical energy used to operate the refrigeration system.

The system absorbs energy from the heat source (*net capacity* of the system) plus the heat of the electrical energy used to absorb the heat (*motor heat*). The net capacity plus the motor heat equal the heat energy that goes into the heat sink (*gross capacity* of the system). In other words, the *heat removed* plus the *heat used to remove that heat* equal the gross capacity. The basic formula for this function is as follows:

Gross capacity = Net capacity + Motor heat

When the system is operating, there are heat losses from those portions of the piping system that have higher temperatures than the atmospheric temperatures surrounding the piping. There are also heat gains from those portions of the piping that have lower temperatures than the atmospheric temperatures surrounding the piping. These losses and gains will be discussed in more detail later in the text. For this portion of the text, it will be assumed that the net capacity plus the motor heat equal 100% of the gross capacity.

It has not been stated what percentage of the gross capacity constitutes the net capacity and the motor heat. The objective of fine tuning is to produce as high a percentage of net capacity as possible using as low a percentage of motor heat as possible. The balance between the net capacity plus the motor heat and the gross capacity is known as the *thermal balance* of the system. **The amount of heat energy *in* will always equal the amount of heat energy *out*.**

Factors That Affect Comfort

An air conditioner or heat pump is supposed to remove enough sensible heat and moisture to maintain the most comfortable temperature and humidity in the conditioned area.

In the cooling mode, the air conditioning system or heat pump draws air through the evaporator, treating the air from the conditioned area as its heat source. Heat in this area is composed of heat that affects the temperature of the air as well as the moisture in the air. The heat that affects the temperature of the air, the *sensible heat* (Hs), is brought into the area by transmission through the building structure; air leakage, also known as infiltration; and internal sources such as people, lighting, and equipment. Moisture, the *latent heat* (Hl) portion of the cooling load, is brought by conduction through the building construction as well as internal sources.

In the early days of comfort conditioning, experiments were conducted by having people occupy an area where conditions of temperature and humidity could be varied and closely controlled. The goal was to determine the conditions in which the greatest number of people were comfortable. The experiments showed that different people are comfortable in a certain range of conditions — in what is known as the *comfort zone* of air conditioning.

Three characteristics in the conditioned area have an effect on people: air movement, temperature, and relative humidity. *Relative humidity*, expressed as a percentage, is the maximum amount of moisture that the air can hold at a certain temperature. If air is moving more than 50 feet per minute (fpm), it produces uncomfortable drafts. The range of movement at which drafts are minimized is 15 to 25 fpm. The temperature range in the comfort tests was from 68° to 85°F. The relative humidity range in the tests was from 30% to 70%.

The results of the tests revealed that comfortable temperatures and relative humidity have a converse relationship. If the humidity is reduced, the maintained temperature can be higher and people can still be comfortable. Conversely, if the humidity is increased, the temperatures must be lowered to produce comfortable conditions.

Each comfortable combination of temperature (*dry bulb temperature*) and humidity (*wet bulb temperature*) became known as the *effective temperature* (et). Dry bulb (db) temperature measures with a dry bulb thermometer the heat intensity of air temperature only, not humidity. Wet bulb (wb) temperature measures with a wet bulb thermometer, the bulb of which is covered with a water-soaked wick. The lowering of temperature that results from the evaporation of water around the bulb indicates the air's relative humidity.

The recommended temperature and humidity for maximum comfort and energy conservation are 80°F dry bulb and 66.5°F wet bulb or 50% relative humidity. The air flowing through the evaporator must be set to maintain these conditions in the conditioned area despite a wide range of outdoor temperatures and relative humidity.

Heat produced in people and other animals is removed from the body by evaporation of moisture from the outer-body surfaces. It is possible to be uncomfortable despite the temperature surrounding the body if the moisture content of the air is extremely high. The amount of moisture in the air has a direct effect on the comfort level in the conditioned area. The uncomfortable feeling of dryness in the winter or the muggy feeling of high humidity in the summer means that the heat emission rate of the occupants is being seriously affected.

Moisture travels from one place to another because of the difference in pressure that the water vapor produces. Water standing in liquid form has higher vapor pressure than moisture in the air at the surface of the water. The vapor pressure of the liquid water also depends upon its temperature. The higher the liquid water sensible temperature (temperature that can be measured with a thermometer), the higher the vapor pressure. This is why hot water vaporizes faster than cold water.

To fine tune an air conditioning system or heat pump, the correct heat quantity input to the evaporator must be established. To do this, it is necessary to determine the ratio of sensible heat and relative humidity in the air to be conditioned.

PSYCHROMETRY

Psychrometry is the science of the characteristics of air, regarding heat and moisture content and specific volume. Knowledge of the basic principles, terms, and the relationship of the elements in the air is necessary to be able to adjust the heat energy input into the air conditioning system or heat pump. The main terms of psychrometry are as follows:

ATMOSPHERE

The air, or atmosphere, surrounding the Earth is made up of a mixture of dry gases and water vapor. The dry gases are approximately 77% nitrogen, 22% oxygen, and 1% other gases such as hydrogen and other rare gases. Water vapor exists in varying quantities depending upon the location of the air and its exposure to water. The amount of moisture in the air is higher next to and over bodies of water than over dry or desert areas. The water quantities in air are so small that a measurement called *grain* is used. There are 7000 grains of moisture in a pound of water.

DRY BULB TEMPERATURE

Dry bulb temperature was discussed previously. Sensible heat is heat energy that, when added to or removed from a substance, causes a rise or fall in temperature. The sensible heat intensity or temperature of the air is measured by a thermometer with a dry surface, which is called a dry bulb thermometer.

WET BULB TEMPERATURE

Wet bulb temperature was also discussed previously, but more explanation is required. Evaporation is used to measure the amount of moisture in the air. A thermometer bulb is wrapped in a wick material saturated with water. Air passes over the wick, causing water to evaporate. The heat is drawn from the wick, lowering its temperature, which causes the bulb to lose heat and the thermometer reading to fall.

There is a limitation to this method, however. When the wick temperature is lowered, a sensible difference in temperature is created between the wick and the air. As the wick temperature drops and the difference in temperature increases, the sensible heat flow to the wick increases. The actual wet bulb temperature reading is the balance point between the difference in temperature resulting from the water evaporation lowering the wick temperature and the sensible temperature raising the wick temperature. The drier the air, the faster the water is evaporated and the lower the balance temperature of the two heat transfers.

The dry bulb and wet bulb readings on the thermometers show the amount of moisture in the air compared to how much it would have if it were completely saturated. The difference between the dry bulb and wet bulb readings is called the *wet bulb depression.*

DEW POINT TEMPERATURE

Air expands or contracts depending on temperature changes. A rise in temperature causes the air to expand; therefore, a pound of air by weight will occupy a greater volume of space if it is heated. A drop in temperature reduces the volume of the pound of air.

If the air is compared to a bucket that can hold water, a comparison of water quantities can be made. For example, if a bucket 12 in. high is filled with water to a height of 6 in., the bucket is half full or the water occupies 50% of the bucket. If the height of the bucket were reduced to 8 in. without any water removal from the bucket, the water now occupies 75% of the bucket. One could say that the relative humidity of the bucket is 75%. The bucket is now holding 75% of the water it could hold if it were filled.

If the height of the bucket is reduced to 6 in. with 6 in. of water in the bucket, the water now occupies 100% of the bucket. The relative humidity of the bucket is 100%. At this point, any further reduction in the height of the bucket would cause water to flow over the top of the bucket. One could say this is the *dew point* (dp) of the bucket.

Because a change in the temperature of the air causes a change in the volume of the air, as the temperature drops, the volume drops.

As the volume drops, the amount of moisture in the air represents a higher percentage of the total volume of air and moisture or a higher relative humidity in the air. If the temperature and volume of the air continue to be reduced, it will reach a saturation point at which it contains all the moisture it can hold. Any further reduction in temperature will cause moisture to drop out of the air. The temperature at which this occurs is called the dew point. In weather terms, water that drops out of the air is called fog. When the moisture becomes excessive, it is called rain.

In air conditioning work, the temperature at which water forms on the evaporator is called the *condensing temperature* of the moisture in the air, and the resulting liquid is called *condensate.*

SPECIFIC HUMIDITY

Specific humidity (SH) is the actual weight of water vapor expressed in grains of water per pound of dry air. It also may be expressed as pounds of water per pound of dry air, depending on the data used compared to how much water the air can hold at saturation. This comparison is without any change in sensible temperature.

RELATIVE HUMIDITY

Relative humidity (RH) is the ratio of water vapor, expressed as a percentage, in air compared to the amount of water vapor the air could hold without a change in temperature.

SPECIFIC VOLUME

Specific volume (Vs) is the number of cubic feet (ft^3) occupied by one pound of the mixture of dry air and water vapor.

SENSIBLE HEAT

Sensible heat (H_s) is the amount of heat in the air, expressed in British thermal units per pound (Btu/lb), that produces the dry bulb temperature of the air. Sensible heat is heat energy that causes a rise or fall in temperature when added to or removed from a substance.

LATENT HEAT

Latent heat (Hl) is the heat required to vaporize the moisture contained in a specific quantity of air. This evaporation, or vaporization, occurs at the wet bulb temperature of the air and is expressed in Btu/lb. Latent heat causes a substance to change state between solid, liquid, and gas rather than just a temperature change.

TOTAL HEAT OR ENTHALPY

The *total heat* (Ht) content of the air is called the *enthalpy* of the air. It is the sum of the sensible heat and the latent heat quantities in Btu/lb of the mixture of dry air and water vapor. The enthalpy of the air is measured at the saturation point or wet bulb temperature of the air.

PSYCHROMETER

To measure the properties of air, a *psychrometer* or wet bulb hygrometer is used. This instrument has two temperature-sensing devices that simultaneously measure the properties of the air being sampled. If the instrument uses a blower to force the air over the thermometers, it is known as a *power-type psychrometer*. These usually are permanently mounted instruments such as those used in industry or in laboratory applications.

The commonly used type of *sling psychrometer* is one that uses two thermometers mounted on a flat plate. It is constructed to have a swivel arrangement to move the instrument through the air. This produces the same effect of air motion contact to the sensing thermometers or elements.

Figure 1-1 shows a type of sling psychrometer that uses two glass thermometers mounted on the flat plate with a swivel

Figure 1-1. Sling psychrometer (Courtesy, Weksler Instruments Corporation)

handle. Rotating the instrument at a speed of two to three revolutions per second exposes the thermometers to the proper airflow for accurate results. The lower thermometer has no wick or exposure to moisture, so it reads the sensible temperature of the air. The upper thermometer has a wet wick. This thermometer reads the effect of the moisture evaporation from the wick versus the sensible temperature difference (TD) between the wick and the air sample. These two temperatures are used to determine the heat and moisture content of the air.

PSYCHROMETRIC CHART

When all the properties that exist in air in the forms of sensible heat and water vapor and the various proportions of the properties are plotted on a chart, the chart can be used to determine heat and moisture content and specific volume. Specific volume is the space occupied by a substance per unit of mass. Figure 1-2 shows a form of psychrometric chart that includes information on the properties of air in various combinations. The scale at the bottom of the chart is the dry bulb temperature of the air. The vertical lines from this scale represent the dry bulb temperatures.

The wet bulb temperatures of the air are plotted as lines that slope downward from left to right. The chart also shows the relative humidity at the various combinations of dry bulb and wet bulb temperatures. The lowest curved line shows 10% relative humidity, and each line above it represents a 10% increase. The lines that slope upward to the left represent the specific volume of air in cubic feet per pound of dry air at the various combinations of air and moisture. The range of specific volumes for comfort conditions is from 12.3 to 15.0 cubic feet per pound.

These are the basic factors of the psychrometric chart and are intended to show the relationship between the factors. For example, suppose that the psychrometer is used to measure the dry bulb and wet bulb temperatures in the return air of an air conditioning system, and suppose that the dry bulb temperature of the air is 80°F and the wet bulb temperature is 66.5°F, Figure 1-3. Using Figure 1-3, the point where the vertical 80°F dry bulb line and the angled 66.5°F wet bulb line cross is on the 50% relative humidity line (Point A). Point A is also between the 13.5 and the 14.0 specific volume lines. It is more than halfway between the lines at approximately 13.85 ft^3/lb. This indicates that the specific volume of the air and moisture mixture at these conditions is 13.85 ft^3/lb of dry air.

To show the effect of a change in the sensible heat content of the air, suppose the dry bulb temperature of the air is increased to 90°F without adding or removing any moisture (from Point A to Point B, Figure 1-3). The following changes take place:

- The volume of the air increases from 13.85 ft^3/lb to approximately 14.08 ft^3/lb.

- Because the air expands and its ability to hold water increases (the bucket gets larger) the percent of moisture in the air decreases. The air-to-moisture ratio drops from 50% to 37% relative humidity.

- The wet bulb temperature increases to 69.6°F because of the increase in sensible heat added to the wet bulb wick. The higher sensible heat and the small increase in the evaporation rate of the water from the wick results in a higher temperature balance or wet bulb temperature reading.

Suppose instead of adding sensible heat and raising the dry bulb temperature, the dry bulb temperature is kept constant and only moisture is added to the air. To vaporize the water requires latent heat energy — 980 Btu/lb of water. This latent heat is added to the total amount of heat in the air without affecting the dry bulb temperature of the air. Adding the moisture, however, raises the relative humidity of the air and the wet bulb temperature. Assume that sufficient moisture is added to the air to raise the relative humidity to 70% (Point A to Point C, Figure 1-3). This reduces the evaporation effect on the wick of the wet bulb thermometer. As a result, the wick temperature rises, and a new temperature balance is found between the evaporation heat loss from the wick and the sensible heat gain, which results in a wet bulb temperature reading of 72.5°F.

If moisture is added to the air until it is completely saturated (100% relative humidity) (Point D, Figure 1-3), the evaporation loss from the wick is zero and the wick registers the dry bulb temperatures only. The wet bulb and dry bulb temperatures are the same at 100% relative humidity.

It has been shown that a rise in dry bulb temperature raises the specific volume and

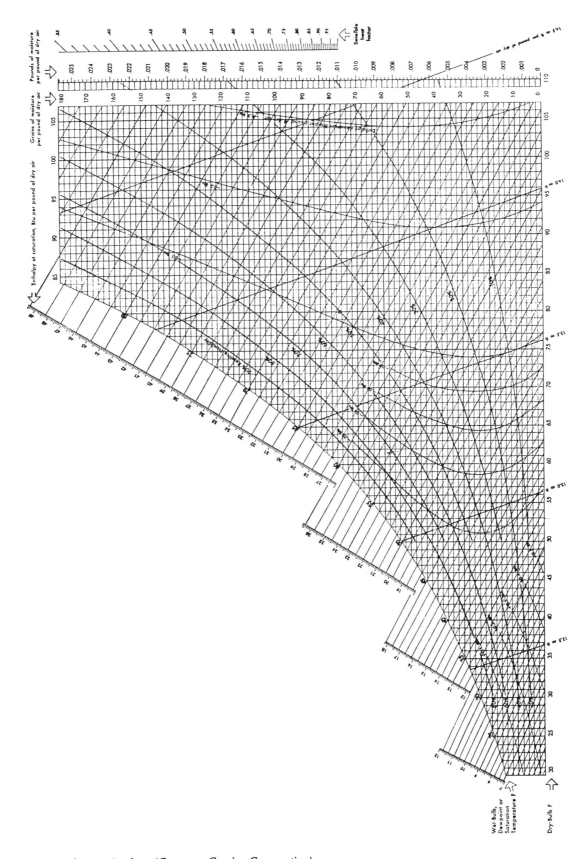

Figure 1-2. Psychrometric chart (Courtesy, Carrier Corporation)

Thermal Balance and Psychrometrics 7

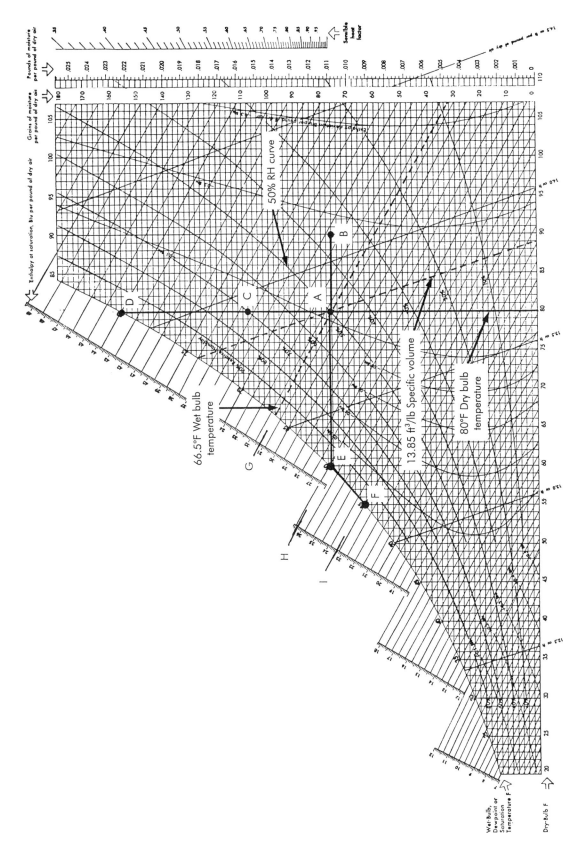

Figure 1-3. Psychrometric chart with plot points (Courtesy, Carrier Corporation)

the wet bulb temperature and lowers the relative humidity. If the procedure is reversed and the dry bulb temperature of the air is lowered, the wet bulb temperature and specific volume decrease, while the relative humidity increases. For example, if the dry bulb temperature of the air is lowered to 59.5°F (Point A to Point E, Figure 1-3), the specific volume decreases to approximately 13.31 ft³/lb and the relative humidity increases to 100%. This point is called the dew point of moisture in air or the condensing temperature of other vapors such as refrigerants.

The air cannot hold more moisture after it is 100% saturated. Therefore, any further reduction in the sensible temperature of the air results in condensation of the moisture from the air. Suppose the dry bulb temperature of the air is reduced to 55°F (Point F, Figure 1-3). The wet bulb temperature is the same as the dry bulb temperature, because the relative humidity is still 100%. The specific volume drops to 13.15 ft³/lb.

Up to this point, only the addition or deletion of sensible and latent heat have been discussed. No attempt has been made to determine the amount of sensible or latent heat involved in the changes.

TOTAL HEAT IN AIR

As mentioned previously, the total heat quantity in air is the sum of the sensible heat and the latent heat quantities in Btu/lb of the mixture of dry air and water vapor. Total heat is also called enthalpy. The enthalpy of the air is measured at the saturation point or wet bulb temperature of the air.

Consider the original 80°F dry bulb temperature and 50% relative humidity conditions of the air (Point A, Figure 1-3), and follow the wet bulb line to the enthalpy scale. The total heat content of the air is 31.3 Btu/lb (Point G, Figure 1-3). When the air is cooled to 59.5°F, the total heat content at that temperature is 26.4 Btu/lb (Point H, Figure 1-3). Therefore, a total quantity of heat of 4.9 Btu/lb (31.3 Btu/lb - 26.4 Btu/lb) is removed. The results using either the sensible heat scale or the enthalpy scale are the same, because the amount of moisture in the air did not change.

When latent heat is part of the heat quantity removed, only the total heat or enthalpy scale can be used. For example, when the temperature of the air is lowered to 55°F, sensible and latent heat are removed. The air is at a dry bulb temperature of 80°F and 50% relative humidity containing enthalpy of 31.3 Btu/lb. At a dry bulb temperature of 55°F, the enthalpy is 23.4 Btu/lb (Point I, Figure 1-3). What is removed is an enthalpy of 7.9 Btu/lb (31.3 Btu/lb - 23.4 Btu/lb). The difference between the enthalpy of 7.9 Btu/lb and the sensible heat of 6.05 Btu/lb is 1.85 Btu/lb of latent heat of moisture removed.

To determine the actual quantity of condensate, or water, removed from the air, the scale on the right side of the psychrometric chart is used. This scale is in grains of moisture per pound of air. The air at Point A, Figure 1-3, contains 75 grains of moisture per pound of air. The grains of moisture per pound of air remain the same from Point A to Point E, Figure 1-3. However, from Point E to Point F, water condenses out of the air. At 55°F, the air contains 64 grains of moisture per pound of air. From Point E to Point F, 11 grains of moisture per pound are removed from the air (75 grains/lb - 64 grains/lb).

All the factors involved in the psychrometric chart will be used in further discussions of system capacity calculations and operating efficiencies.

PROBLEMS

1.1. Define thermal balance.

1.2. Define enthalpy.

1.3. Define relative humidity.

1.4. Define psychrometry.

1.5. Define dry bulb temperature.

1.6. Define wet bulb temperature.

1.7. Define dew point.

1.8. Define condensing temperature.

1.9. Define specific humidity.

1.10. Define sensible heat.

1.11. Define latent heat.

1.12. Define specific volume.

1.13. Match the item with the abbreviation.

1. Relative humidity	A. et
2. Dew point	B. Vs
3. Effective temperature	C. RH
4. Specific humidity	D. wb
5. Dry bulb temperature	E. SH
6. Enthalpy	F. dp
7. Specific volume	G. Hl
8. Latent heat	H. db
9. Wet bulb temperature	I. Hs
10. Sensible heat	J. Ht

CHAPTER TWO
Adjusting Airflow

Two items determine the input load on an air conditioning system or heat pump — the heat content of the air and the amount of air flowing through the evaporator. The heat content of the air was discussed in Chapter 1. This chapter will focus on the proper amount of air flowing through the evaporator.

The amount of air passing through the indoor evaporator when operating in the cooling mode may not be the same for checking the system as it would be for obtaining the proper conditions in the conditioned area. The manufacturer-recommended rate in cfm for design conditions and testing must be used to determine if the unit is working properly.

Units are tested and rated by the Air Conditioning and Refrigeration Institute (ARI) at these standard conditions:

- 95°F outdoor ambient temperature
- 80°F indoor dry bulb temperature
- 66.5°F indoor wet bulb temperature
- 50% relative humidity (RH)
- 400 cfm per 12,000 Btuh unit cooling capacity

Throughout this chapter, the ARI standard of 400 cfm per 12,000 Btuh unit cooling capacity will be used to determine the desired change in temperature for the test unit. Some manufacturers do not strictly follow this standard. For the greatest accuracy in the test procedure, use the cfm quantities given in the manufacturer's specification. It is imperative that the service technician obtain and retain the performance specifications on all products possible in order to produce peak system performance. Manufacturers do not retain literature on discontinued products.

To determine how much air should be used to test the unit at these standard conditions, it is only necessary to divide the gross (total) unit capacity at 95°F outdoor ambient temperature by 12,000 Btuh and multiply this ratio by 400 cfm. For example, the 30HPQ4 unit in Figure 2-1 is rated at 29,200 Btuh at 95°F:

$$cfm = \frac{(Btuh)(400 \text{ cfm})}{12,000 \text{ Btuh}}$$

$$cfm = \frac{(29,200 \text{ Btuh})(400 \text{ cfm})}{12,000 \text{ Btuh}} = 973 \text{ cfm}$$

This is the desired rate in cfm that should be flowing through the evaporator.

The heating unit is used as a basis for determining the cooling cfm. Once the cfm

HEATING APPLICATION DATA RATINGS — Outdoor Temperature °F*

OUTDOOR MODEL	INDOOR MODEL		0°	5°	10°	15°	17°	20°	25°	30°	35°	40°	45°	47°	50°	55°	60°	65°
18HPQ2	H18QS1	Btuh	6,100	6,900	7,900	8,850	9,300	9,850	10,900	11,800	12,750	14,700	16,700	17,600	18,300	19,500	20,800	22,000
	&	Watts	1,750	1,780	1,810	1,840	1,850	1,865	1,885	1,900	1,930	1,980	2,030	2,050	2,080	2,135	2,185	2,240
	B18EHQ1	COP	1.02	1.14	1.28	1.41	1.47	1.55	1.69	1.82	1.94	2.18	2.41	2.52	2.58	2.68	2.79	2.88
18HPQ2	H24QS1	Btuh	6,700	7,500	8,400	9,300	9,625	10,200	11,100	12,000	12,800	14,600	16,300	17,000	17,700	18,800	20,000	21,100
	&	Watts	1,675	1,700	1,725	1,735	1,750	1,765	1,775	1,790	1,800	1,850	1,870	1,880	1,890	1,925	1,950	1,975
	B24EHQ1	COP	1.17	1.29	1.43	1.57	1.61	1.69	1.83	1.96	2.08	2.31	2.55	2.65	2.74	2.86	3.01	3.13
24HPQ2	H18QS1	Btuh	9,800	10,400	11,500	12,200	12,425	12,900	13,600	14,200	14,900	18,300	21,700	23,200	24,100	25,700	27,300	29,000
	&	Watts	1,900	2,000	2,075	2,150	2,200	2,250	2,325	2,425	2,470	2,575	2,925	2,725	2,750	2,850	2,925	3,025
	B18EHQ1	COP	1.51	1.52	1.62	1.66	1.65	1.68	1.71	1.72	1.77	2.08	2.17	2.50	2.57	2.64	2.73	2.81
24HPQ2	H24QS1	Btuh	8,600	9,900	11,300	12,700	13,300	14,200	15,500	16,900	18,200	20,700	23,000	24,000	25,000	26,600	28,200	30,000
	&	Watts	2,060	2,110	2,165	2,225	2,250	2,280	2,340	2,390	2,450	2,535	2,625	2,660	2,710	2,790	2,870	2,950
	B24EHQ1	COP	1.22	1.37	1.53	1.67	1.73	1.82	1.94	2.07	2.18	2.39	2.57	2.65	2.70	2.79	2.88	2.98
30HPQ4	H3AQ1	Btuh	13,000	14,800	16,500	18,300	18,950	20,000	21,700	23,400	25,000	27,900	30,900	32,000	33,250	35,250	37,200	39,200
		Watts	2,980	3,010	3,040	3,060	3,095	3,110	3,120	3,150	3,200	3,260	3,840	3,395	3,440	3,540	3,650	3,770
		COP	1.28	1.44	1.59	1.75	1.79	1.88	2.04	2.18	2.29	2.51	2.71	2.75	2.83	2.92	2.99	3.05
30HPQ4	B30EHQ	Btuh	11,400	13,300	15,200	17,100	17,800	19,000	20,900	22,800	24,700	27,600	30,400	31,600	33,000	35,300	37,600	39,700
		Watts	2,550	2,650	2,750	2,850	2,890	2,950	3,050	3,150	3,240	3,365	3,485	3,530	3,600	3,715	3,830	3,945
		COP	1.30	1.45	1.60	1.75	1.80	1.90	2.00	2.10	2.25	2.40	2.55	2.60	2.70	2.80	2.85	2.95
30HPQ4	B36EHQ1	Btuh	12,600	14,300	16,000	17,700	18,400	19,400	21,100	22,800	24,400	27,600	30,700	32,000	33,300	35,400	37,500	39,500
		Watts	2,690	2,765	2,845	2,920	2,960	2,995	3,070	3,145	3,200	3,295	3,370	3,405	3,440	3,515	3,590	3,665
		COP	1.37	1.52	1.65	1.78	1.82	1.90	2.00	2.12	2.23	2.45	2.67	2.75	2.84	2.95	3.06	3.16
36HPQ4	H3AQ1	Btuh	16,300	18,300	20,200	22,400	23,200	24,500	26,700	28,700	30,850	34,800	38,700	40,500	42,000	44,600	47,300	49,800
		Watts	3,200	3,330	3,460	3,600	3,665	3,730	3,860	3,990	4,080	4,250	4,390	4,450	4,520	4,650	4,790	4,920
		COP	1.49	1.61	1.71	1.82	1.85	1.92	2.03	2.11	2.22	2.40	2.58	2.65	2.72	2.81	2.89	2.97
36HPQ4	B36EHQ1	Btuh	14,200	16,500	19,000	21,400	22,500	23,900	26,300	28,700	31,200	35,300	39,300	41,000	42,600	45,400	48,200	51,000
		Watts	3,200	3,320	3,440	3,550	3,600	3,670	3,790	3,910	4,020	4,210	4,390	4,475	4,570	4,750	4,940	5,130
		COP	1.30	1.46	1.62	1.77	1.83	1.91	2.03	2.15	2.27	2.46	2.62	2.70	2.73	2.80	2.86	2.91
42HPQ	H4AQ1	Btuh	14,000	16,200	18,800	21,400	22,300	24,100	26,800	29,400	32,000	36,000	40,000	41,500	43,200	46,200	49,000	52,000
		Watts	3,530	3,640	3,740	3,840	3,878	3,940	4,040	4,140	4,246	4,340	4,440	4,475	4,530	4,630	4,720	4,810
		COP	1.16	1.31	1.47	1.63	1.68	1.79	1.94	2.08	2.20	2.43	2.64	2.70	2.79	2.92	3.04	3.17
42HPQ	BC48A	Btuh	16,200	18,100	20,000	22,000	22,700	23,900	25,800	27,800	29,600	34,200	38,600	40,500	42,000	44,700	47,400	50,100
		Watts	3,660	3,760	3,860	3,970	4,010	4,070	4,180	4,280	4,385	4,490	4,590	4,630	4,690	4,800	4,900	5,010
		COP	1.29	1.41	1.51	1.62	1.66	1.72	1.81	1.91	1.98	2.23	2.46	2.55	2.62	2.72	2.83	2.93
48HPQ2	H4AQ1	Btuh	18,000	19,800	22,000	24,400	25,100	26,700	29,000	31,400	33,750	38,000	42,200	44,000	45,600	48,400	51,300	54,200
		Watts	3,760	3,900	4,030	4,160	4,210	4,290	4,440	4,540	4,660	4,770	4,880	4,914	4,970	5,060	5,130	5,180
		COP	1.40	1.48	1.61	1.70	1.74	1.82	1.91	2.03	2.12	2.28	2.53	2.60	2.69	2.80	2.93	3.07
48HPQ2	BC48A	Btuh	17,600	19,200	21,600	24,200	25,200	26,800	29,400	31,900	34,400	38,700	42,800	44,500	46,200	49,100	52,000	55,000
		Watts	3,840	4,000	4,150	4,310	4,368	4,460	4,590	4,720	4,826	4,910	4,980	4,997	5,030	5,070	5,110	5,150
		COP	1.29	1.40	1.52	1.64	1.69	1.76	1.88	1.98	2.08	2.30	2.52	2.60	2.69	2.83	2.98	3.12
60HPQ4	H5AQ1	Btuh	21,000	24,200	27,500	30,700	31,850	33,800	37,000	40,300	43,600	49,400	55,200	57,500	59,800	63,700	67,500	71,500
		Watts	4,310	4,480	4,650	4,815	4,870	4,975	5,150	5,335	5,520	5,675	5,835	5,890	5,990	6,150	6,315	6,470
		COP	1.43	1.58	1.73	1.87	1.92	1.99	2.11	2.21	2.31	2.55	2.77	2.85	2.93	3.03	3.13	3.24
60HPQ4	BC60A	Btuh	22,000	25,000	28,200	31,400	32,650	34,600	37,800	41,000	44,200	50,000	55,700	58,000	60,200	64,000	67,800	71,600
		Watts	4,520	4,685	4,845	5,000	5,075	5,165	5,325	5,485	5,650	5,825	6,000	6,075	6,185	6,360	6,540	6,710
		COP	1.43	1.56	1.71	1.84	1.88	1.96	2.08	2.19	2.29	2.51	2.72	2.80	2.85	2.95	3.04	3.13

*70°F DB Indoor Return Air at Rated CFM

COOLING APPLICATION DATA — Outdoor Temperature °F*

OUTDOOR MODEL	INDOOR MODEL		70°	75°	80°	85°	90°	95°	100°	105°	110°	115°
18HPQ2	H18QS1	Total Btuh	18,250	18,000	17,750	17,600	17,150	16,700	16,350	15,750	14,900	13,350
	&	Sensible Btuh	11,900	12,000	12,100	12,200	12,250	12,150	12,150	11,950	11,550	10,750
	B18EHQ1	Latent Btuh	6,350	6,000	5,650	5,400	4,900	4,550	4,200	3,800	3,350	2,600
18HPQ2	H24QS1	Total Btuh	20,250	20,200	19,800	19,100	18,300	17,400	16,450	15,450	14,450	13,350
	&	Sensible Btuh	15,350	14,700	14,200	13,750	13,350	13,050	12,700	12,400	12,150	11,900
	B24EHQ1	Latent Btuh	4,900	5,500	5,600	5,350	4,950	4,350	3,750	3,050	2,300	1,450
24HPQ2	H18QS1	Total Btuh	21,900	21,875	21,850	21,800	21,800	21,800	21,000	19,900	18,600	17,100
	&	Sensible Btuh	13,000	13,250	13,450	13,750	14,000	14,300	14,200	14,100	13,900	13,500
	B18EHQ1	Latent Btuh	8,900	8,625	8,400	8,050	7,800	7,500	6,800	5,800	4,700	3,600
24HPQ2	H24QS1	Total Btuh	24,850	24,600	24,300	24,000	23,600	23,200	22,400	21,450	20,350	18,500
	&	Sensible Btuh	15,700	16,000	16,300	16,600	16,800	17,000	17,000	16,900	16,500	15,250
	B24EHQ1	Latent Btuh	9,150	8,600	8,000	7,400	6,800	6,200	5,400	4,550	3,850	3,250
30HPQ4	H3AQ1	Total Btuh	34,400	33,400	32,400	31,400	30,400	29,400	28,300	27,250	26,250	25,250
		Sensible Btuh	22,500	22,350	22,200	22,100	21,950	21,800	21,700	21,550	21,400	21,250
		Latent Btuh	11,900	11,050	10,200	9,300	8,450	7,600	6,600	5,700	4,850	4,000
30HPQ4	B30EHQ	Total Btuh	30,700	30,250	29,750	29,150	28,500	27,800	26,750	25,550	24,350	22,900
		Sensible Btuh	19,650	19,600	19,400	19,150	18,850	18,500	18,100	17,600	17,100	16,500
		Latent Btuh	11,050	10,650	10,350	10,000	9,650	9,300	8,650	7,950	7,250	6,400
30HPQ4	B36EHQ1	Total Btuh	31,750	31,600	31,300	30,850	30,150	29,200	28,000	26,700	25,250	23,650
		Sensible Btuh	22,300	22,250	22,200	22,100	21,800	21,600	21,100	20,600	20,100	19,500
		Latent Btuh	9,450	9,350	9,100	8,750	8,350	7,600	6,900	6,100	5,150	4,100
36HPQ4	H3AQ1	Total Btuh	40,200	39,500	38,800	38,050	37,350	36,600	35,600	34,300	32,800	31,000
		Sensible Btuh	26,500	26,200	25,950	25,750	25,600	25,550	25,400	25,300	25,200	25,200
		Latent Btuh	13,700	13,300	12,850	12,300	11,750	11,000	10,200	9,000	7,600	5,800
36HPQ4	B36EHQ1	Total Btuh	38,700	38,550	38,200	37,750	37,000	35,800	34,300	32,500	30,350	27,900
		Sensible Btuh	25,900	26,050	26,150	26,250	26,300	26,300	25,850	25,200	24,250	23,050
		Latent Btuh	12,800	12,500	12,050	11,500	10,700	9,500	8,450	7,300	6,100	4,850
42HPQ	H4AQ1	Total Btuh	50,100	48,700	47,200	45,600	44,000	42,500	40,800	39,100	37,400	35,600
		Sensible Btuh	35,600	35,300	35,000	34,600	34,200	33,700	33,300	32,800	32,200	31,700
		Latent Btuh	14,500	13,400	12,200	11,000	9,800	8,800	7,500	6,300	5,200	3,900
42HPQ	BC48A	Total Btuh	46,600	45,800	45,000	44,000	42,800	41,500	40,000	38,400	36,800	35,000
		Sensible Btuh	33,000	32,800	32,600	32,400	32,000	31,700	31,400	31,000	30,600	30,200
		Latent Btuh	13,600	13,000	12,400	11,600	10,800	9,000	8,000	7,400	6,200	4,800
48HPQ2	H4AQ1	Total Btuh	50,700	50,200	49,600	49,100	48,000	46,500	44,700	42,800	40,700	38,500
		Sensible Btuh	36,100	36,000	35,800	35,700	35,200	34,700	34,200	33,800	33,200	32,600
		Latent Btuh	14,600	14,200	13,800	13,400	12,800	11,800	10,500	9,000	7,500	5,900
48HPQ2	BC48A	Total Btuh	51,100	50,400	49,400	48,400	47,000	45,500	43,700	41,800	39,900	37,800
		Sensible Btuh	34,700	34,600	34,400	34,200	34,000	33,600	33,400	33,200	32,800	32,500
		Latent Btuh	16,400	15,800	15,000	14,200	13,000	11,900	10,300	8,600	7,100	5,300
60HPQ4	H5AQ1	Total Btuh	62,500	61,200	60,100	59,200	58,400	57,500	55,900	53,450	50,350	46,500
		Sensible Btuh	39,170	39,300	39,350	39,450	39,475	39,360	39,100	38,500	37,550	36,000
		Latent Btuh	23,330	21,900	20,750	19,750	18,925	18,150	16,800	14,950	12,800	10,500
60HPQ4	BC60A	Total Btuh	62,500	61,300	60,100	59,000	58,000	57,000	55,000	52,500	49,500	45,325
		Sensible Btuh	37,700	38,200	38,600	38,975	39,300	39,450	38,675	37,775	36,500	35,100
		Latent Btuh	24,800	23,100	21,500	20,025	18,700	17,550	16,325	14,725	13,000	10,225

*At 80°DB/67°WB Return Air Temperature at Rated Indoor CFM.

*At 80° DB/67° WB Return Air Temperature at Rated Indoor CFM.

Figure 2-1. Manufacturer specifications (Courtesy, Bard Manufacturing Co.)

is known, it is possible to determine the cooling capacity. To determine the amount of heat a unit is adding (heating) or removing (cooling) from the area supplied to a conditioned area, it is necessary to know the amount of air moved (in cfm) and the amount of heat added to or removed from each cubic foot.

There are several methods by which the desired airflow can be obtained. These are:

* Air temperature change

* Air duct velocity

* External static pressure (ESP) of the air handler

* Total cubic feet per minute measurement

AIR TEMPERATURE CHANGE

The difference in temperature (TD) method requires using the heating unit of a heating/cooling system or heat pump that has a measurable heat energy source. This method also requires that the user know the efficiency of the unit that will convert the heat energy from the source into heat energy in the air. The heat energy sources in present-day equipment are electricity and fossil fuels (natural gas, liquefied petroleum gas, otherwise known as LP gas, and oil).

Whatever heat energy source is used, the goal is to find the difference in temperature that will result from the source output and the cfm required by the cooling unit at standard conditions of 400 cfm per 12,000 Btuh cooling unit capacity. This temperature change must be determined before the heating unit can be adjusted for the desired cfm. This method is different from adjustment practices used in the past, because the cfm required for the cooling unit operation is known. The output of the heating unit is determined from the input in British thermal units per hour (Btuh) times the efficiency of the unit. The desired temperature change through the heating unit can be determined using the following formula:

$$TD = \frac{(Btuh\ output)(Efficiency)}{(Cooling\ cfm)(1.08)}$$

where: TD = difference in temperature (°F) the heat energy produces in the air as it passes through the heating unit. For the greatest accuracy, the temperature of the air entering the unit (*return air*) and the air leaving the unit (*supply air*) are measured at the heating unit. This eliminates the effect of heat gain or loss in the air distribution system.

Btuh = amount of heat in Btuh that the heating unit transfers into the air.

1.08 = a constant used to convert cubic feet per minute to cubic feet per pound, and from pounds per minute to pounds per hour. It also contains the quantity of heat energy needed to change the temperature of each pound of air. This figure is derived from the following formula:

$1.08 = (60\ min/hr)(0.075\ lb/ft^3)(0.24\ Btu/lb)$

To convert British thermal units per hour to British thermal units per minute, divide the British thermal units per hour by 60. Cubic feet of air cannot be used to determine the heat content, because air expands or contracts with temperature change. However, a pound of air is a pound of air regardless of its temperature and/or humidity content and how much space it occupies.

Before the difference in temperature method can be calculated, it is first necessary to determine the Btuh input of various units, as well as the efficiencies of these units.

Electric Units

The efficiency of a resistance-type heater element is 100%. The amount of heat energy equivalent to the electrical energy going into the element will result in the same amount of heat energy going from the element into the air over it. Therefore, to measure the amount of heat energy coming from the element, it is only necessary to determine the watts of electrical energy the element is using. The watts of electrical energy in the element per hour multiplied by 3.413 Btu/watt are used to determine the heat energy in Btuh.

To determine the watts per hour flowing into the element, the voltage applied to the element and the amperage flowing through the element are measured simultaneously. The following formula is used:

watts = (volts, V)(amperes, A)

When the two formulas are combined, the formula becomes:

Btuh = (V)(A)(3.413 Btu/watt)(Efficiency)

The efficiency factor must be considered when figuring the output of the heating unit. The efficiency of an electric element is 100%, so the efficiency factor in the previous formula is 1.0. If a fossil fuel-powered heating unit is used, the efficiency of the unit will be less than 100% and the efficiency factor will be less than 1.0. This will be discussed in detail later in this chapter.

Example 2-1. An electric heating unit has 240 volts applied to it, and the total amperage drawn by the unit is 20 amperes. What is the output in Btuh?

Solution 2-1. The formula used in this calculation is:

Btuh = (V)(A)(3.413 Btu/watt)(Efficiency)
Btuh = (240 V)(20 A)(3.413 Btu/watt)(1.0)
Btuh = 16,382.40 Btuh

This calculation shows that the heating unit is adding 16,382.40 Btuh to the air.

To adjust the blower output to obtain the desired temperature increase, it is necessary to know the temperature increase in the air resulting from the amount of heat added. To find this rate for 400 cfm per 12,000 Btuh, use the cooling unit capacity at 95°F outdoor ambient. For the example used earlier in the chapter, it was determined that 973 cfm through the evaporator is needed. If another unit is used, a different cfm may result.

The electric heating unit has an input of 16,382.40 Btuh and is 100% efficient. Therefore, the output is 16,382.40 Btuh x 1.0 or 16,382.40 Btuh. Using the earlier formula, find the desired TD as follows:

$$TD = \frac{(Btuh\ output)(Efficiency)}{(Cooling\ cfm)(1.08)}$$

$$TD = \frac{(16,382.4\ Btuh)(1)}{(973\ cfm)(1.08)} = 15.6°F$$

With the heating unit operating, adjust the cfm to produce the 15.6°F temperature increase. If the resulting temperature increase is less than what is desired, reduce the air quantity. If the resulting temperature increase is higher than what is desired, increase the air quantity. When the resulting temperature increase and desired temperature increase are the same, the airflow in cfm flowing through the unit is correct for testing.

Fossil Fuel-Powered Units

To determine the difference in temperature in a unit using a fossil fuel as the auxiliary heat, the output of the unit must be found along with the temperature increase. The following instruments are needed to determine the output of a fossil fuel-powered heating unit:

- **Two air temperature thermometers** to record the temperature increase by determining the return and supply air temperatures in °F.

- A **high-stack temperature thermometer** to measure the temperature of the products of combustion (*flue products*) coming from the unit's heat exchanger. This thermometer indicates the amount of waste heat from the unit.

- A **CO_2 analyzer** to measure the carbon dioxide (CO_2) in the flue products to determine the efficiency of the combustion process in the unit.

- An **oil pressure gauge** to ensure that the nozzle pressure is set according to the manufacturer's specifications.

- A **stopwatch** for timing the natural gas meter to establish the correct gas quantity flowing through the unit.

- A **water manometer** to establish the correct manifold pressure in LP gas units.

- A **combustion efficiency and stack loss calculator**, a slide rule that uses the CO_2 percentage and stack temperature to determine the efficiency of the unit.

To determine the output of a fossil fuel-powered unit, obtain the amount in Btu per cubic foot of the fuel supplied to the heating unit. Contact the supplying utility for the amount if natural or mixed gas is used. If propane, butane, or a propane-butane mixture is used, contact the local LP gas supplier. The heat content of a gallon of No. 2 fuel oil is assumed to be 140,000 Btu/gallon. Make sure the input in Btuh to the unit agrees with the manufacturer's rating.

Contact the local utility to find out the heat content in Btu per cubic foot of the natural or mixed gas supplied to the gas distribution system. If there is more than one gas utility in the area, contact each one for the information. Not all utilities supply the same type of gas or heat content in their product.

Follow these steps to determine the input to the unit:

- Make sure the heating unit is operating.

- Time the gas supply meter using the Btu per cubic foot of the gas and the flow rate (meter dial rotation time). *Caution: All other gas loads except pilot lights must be off.*

- To determine the required rate at which gas should pass through the meter, calculate the cubic feet of gas that must be burned per hour. This is done by using the following formula:

$$ft^3/hour = \frac{Btuh\ input}{Btu/ft^3}$$

To use this formula, consider a gas-fired heating unit that is rated by the manufacturer at 120,000 Btuh and the utility company is supplying natural gas that is burned at 1050 Btu/ft^3:

$$ft^3/hour = \frac{120,000\ Btuh}{1050\ Btu/ft^3} = 114.28\ ft^3/hr$$

The answer shown here is the amount of gas the furnace is burning per hour. To determine the time the meter test dial will take to make one revolution, reduce the volume per hour to the volume of the test dial.

If the meter has a one minute per revolution test dial, reduce the ft^3 of gas per hour to ft^3 per minute:

$$ft^3/min = \frac{ft^3/hr}{60}$$

Using the number from the previous example:

$$ft^3/min = \frac{114.28\ ft^3/hr}{60} = 1.9$$

This means that 1.9 ft³ of gas should pass through the 1 ft³ per minute dial. Obviously, the dial will travel faster than one revolution per minute. To determine the time in seconds, reverse the cubic feet of gas per minute to minutes per cubic foot and multiply by 60 seconds per minute:

$$\text{second/ft}^3 = \frac{(1 \text{ min})(60 \text{ seconds})}{1.9} = 31.6$$

In the example, to obtain the 120,000 Btuh input from the utility-supplied gas with a heat content of 1050 Btu/ft³, the 1-ft³ dial must complete one revolution in 31.6 seconds. If the time is less, the unit is overfiring and the manifold pressure must be reduced. *Note: Do not change the unit regulator setting more than 1/4 revolution per adjustment.* The input must be set no less than 10% below the manufacturer's rating (-10% to +0%). To maintain equipment life and prevent control problems, the unit must not be overfired. Some manufacturers supply pocket cards with tables that convert the ft³ of gas per hour to meter dial rotation time.

Figure 2-2 is a chart that lists the seconds required per revolution in the left column and cubic feet per hour in the columns under five sizes of meter test dials. In the previously mentioned example, it was determined that the meter will supply 114.28 ft³ of gas per hour and has a 1-cfm test dial.

In Figure 2-2, look down the column under the 1 ft³ test dial to 113 ft³ per hour in the right portion of the chart and then over to the seconds per revolution column. The chart shows that the time required per revolution of the test dial is 32 seconds. The calculated time of 31.6 seconds is more accurate, but the chart time is on the safe side and will result in the accepted input (no less than 10% below manufacturer's rating). If, when using the chart, the required flow rate and chart flow rate do not match, always use the chart flow rate that is lower than the required flow rate. *Do not overfire the heating unit.*

The unit in this example is assumed to be between sea level and 2000 feet above sea level. At altitudes above 2000 feet above sea level, the unit input must be reduced by 4% for each 1000 feet above sea level. If the unit in the example were at 5000 feet, the unit rating would be 120,000 Btuh less 4% times 5. This would be a rating of 120,000 Btuh minus 24,000 Btuh, or a unit rating of 96,000 Btuh. Therefore, the unit would require 96,000 Btuh divided by 1050 Btu/ft³ or 91.42 ft³ of gas per hour. Using the chart in Figure 2-2, the time per revolution of the 1 ft³ test dial is 40 seconds.

The input of an LP gas heating unit is set by the manifold pressure. Using the u-tube manometer, the manifold pressure is set at 11 in. water column (wc). To determine the actual input, the sizes of the main burner orifices are measured with a numbered drill set.

Figure 2-3 shows the input in Btuh using propane, butane, or butane-propane mixtures. These are all based on a manifold pressure of 11 in. wc.

To find the drill size for the 120,000 Btuh input heating unit, divide the rated input by the number of burner orifices in the unit. If, for example, there are five main burner orifices, the 120,000 Btuh total would be divided by five, resulting in 24,000 Btuh to each burner. Using Figure 2-3, for an input of 23,850 Btuh using propane, a No. 54 drill size is needed for the orifice. If butane or a butane-propane mixture is used, the orifice size for the 23,510 Btuh input would be a No. 55 drill size.

The drill sizes in Figure 2-3 are for units installed at sea level. Drill sizes for altitudes above sea level are given in Figure 2-4. After determining the drill size for units at sea level, the drill size for altitudes above sea level can be determined. For example, if a No. 55 drill is required at sea level, at an altitude of 5000 ft, a No. 56 drill size is required.

Seconds for one Revolution	GAS RATE — CUBIC FEET PER HOUR SIZE OF TEST DIAL					Seconds for one Revolution	GAS RATE — CUBIC FEET PER HOUR SIZE OF TEST DIAL				
	¼ cu. ft.	½ cu. ft.	1 cu. ft.	2 cu. ft.	5 cu. ft.		¼ cu. ft.	½ cu. ft.	1 cu. ft.	2 cu. ft.	5 cu. ft.
10	90	180	360	720	1800	50	18	36	72	144	360
11	82	164	327	655	1636	51	..	..	..	141	355
12	75	150	300	600	1500	52	..	..	69	138	346
13	69	138	277	555	1385	53	17	34	..	136	340
14	64	129	257	514	1286	54	..	..	67	133	333
15	60	120	240	480	1200	55	..	..	..	131	327
16	56	113	225	450	1125	56	16	32	64	129	321
17	53	106	212	424	1059	57	..	..	..	126	316
18	50	100	200	400	1000	58	..	31	62	124	310
19	47	95	189	379	947	59	..	..	..	122	305
20	45	90	180	360	900	60	15	30	60	120	300
21	43	86	171	343	857	62	..	..	..	116	290
22	41	82	164	327	818	64	..	..	..	112	281
23	39	78	157	313	783	66	..	..	..	109	273
24	37	75	150	300	750	68	..	..	..	106	265
25	36	72	144	288	720	70	..	..	..	103	257
26	34	69	138	277	692	72	12	25	50	100	250
27	33	67	133	267	667	74	..	..	..	97	243
28	32	64	129	257	643	76	..	..	..	95	237
29	31	62	124	248	621	78	..	..	..	92	231
30	30	60	120	240	600	80	..	..	..	90	225
31	..	..	116	232	581	82	..	..	..	88	220
32	28	56	113	225	563	84	..	..	..	86	214
33	..	..	109	218	545	86	..	..	..	84	209
34	26	53	106	212	529	88	..	..	..	82	205
35	..	..	103	206	514	90	10	20	40	80	200
36	25	50	100	200	500	92	..	..	..	78	196
37	..	..	97	195	486	94	..	..	..	..	192
38	23	47	95	189	474	96	..	..	..	75	188
39	..	..	92	185	462	98	..	..	..	..	184
40	22	45	90	180	450	100	..	..	..	72	180
41	..	..	..	176	439	102	..	..	..	..	178
42	21	43	86	172	429	104	9	17	35	69	173
43	..	..	..	167	419	106	..	..	..	..	170
44	..	41	82	164	409	108	..	..	..	67	167
45	20	40	80	160	400	110	..	..	..	..	164
46	..	..	78	157	391	112	..	..	..	64	161
47	19	38	76	153	383	116	..	..	..	62	155
48	..	..	75	150	375	120	7	15	30	60	150
49	..	..	..	147	367						

Figure 2-2. Gas meter timing chart (Courtesy, Addison Products Company)

Oil

The input to an oil-fired unit will depend upon the capacity of the nozzle(s) and the operating pressure. The nozzle size is stamped on one of the wrench flats of the nozzle in portions of a gallon (e.g., .75 gallons per hour (gph), .85 gph, 1.00 gph, etc.). Physical removal of the firing head is necessary to determine the nozzle size.

The capacity of a standard oil burner nozzle is based on 100 psig (pounds per square inch, gauge) nozzle pressure. Therefore, when running an efficiency test on an oil-fired heating unit, the operating pressure on the nozzle must be 100 psig. If the manufacturer of the heating unit specifies other than 100 psig nozzle pressure, the nozzle capacity curves for the particular heating unit must be used to determine the oil flow rate required.

The input in Btuh to the oil-fired heating unit is based on the standard 140,000 Btu/gallon for No. 2 fuel oil. If, for example, the nozzle used is .85 gph, the input would be 140,000 Btu/gal at 100 psig nozzle pressure times .85 gph, or 119,000 Btuh.

The input of oil-fired heating units must also be reduced at the rate of 4% per 1000 ft of altitude. The nozzle capacity must be changed to produce the required capacity. Reducing the input by reducing the nozzle pressure produces very poor combustion and excessive coking and carbon formation in the heat exchanger.

Determining Unit Efficiency

As stated previously, the difference in temperature (TD) method requires knowing the

LP Gases (Btu per hour at sea level)

	Propane	Butane
Btu per cubic foot	2500	3175
Specific gravity	1.53	2.00
Pressure at orifice (in. wc)	11	11
Orifice coefficient	0.9	0.9

For altitudes above 2000 feet, first select the equivalent orifice size at sea level from Figure 2-4.

Drill Size (Decimal or DMS)	Propane	Butane or Butane-Propane Mixtures
.008	500	554
.009	641	709
.010	791	875
.011	951	1053
.012	1130	1250
80	1430	1590
79	1655	1830
78	2015	2230
77	2545	2815
76	3140	3480
75	3465	3840
74	3985	4410
73	4525	5010
72	4920	5450
71	5320	5900
70	6180	6830
69	6710	7430
68	7560	8370
67	8040	8910
66	8550	9470
65	9630	10,670
64	10,200	11,300
63	10,800	11,900
62	11,360	12,530
61	11,930	13,280
60	12,570	13,840
59	13,220	14,630
58	13,840	15,300
57	14,550	16,090
56	16,990	18,790
55	21,200	23,510
54	23,850	26,300
53	27,790	30,830
52	31,730	35,100
51	35,330	39,400
50	38,500	42,800
49	41,850	45,350
48	45,450	50,300
47	48,400	53,550
46	51,500	57,000
45	52,900	58,500

Figure 2-3. Input for LP gas units and drill sizes for main burner orifices

Equivalent Orifice Sizes at High Altitudes (includes 4% input reduction for each 1000 ft)

Orifice Size at Sea Level	2000	3000	4000	5000	6000	7000	8000	9000	10,000
1	2	2	3	3	4	5	7	8	10
2	3	3	4	5	6	7	9	10	12
3	4	5	7	8	9	10	12	13	15
4	6	7	8	9	11	12	13	14	16
5	7	8	9	10	12	13	14	15	17
6	8	9	10	11	12	13	14	16	17
7	9	10	11	12	13	14	15	16	18
8	10	11	12	13	13	15	16	17	18
9	11	12	12	13	14	16	17	18	19
10	12	13	13	14	15	16	17	18	19
11	13	13	14	15	16	17	18	19	20
12	13	14	15	16	17	17	18	19	20
13	15	15	16	17	18	18	19	20	22
14	16	16	17	18	18	19	20	21	23
15	16	17	17	18	19	20	20	22	24
16	17	18	18	19	19	20	22	23	25
17	18	19	19	20	21	22	23	24	26
18	19	19	20	21	22	23	24	26	27
19	20	20	21	22	23	25	26	27	28
20	22	22	23	24	25	26	27	28	29
21	23	23	24	25	26	27	28	28	29
22	23	24	25	26	27	27	28	29	29
23	25	25	26	27	27	28	29	29	30
24	25	26	27	27	28	28	29	29	30
25	26	27	27	28	28	29	29	30	30
26	27	28	28	28	29	29	30	30	30
27	28	28	29	29	29	30	30	30	31
28	29	29	29	30	30	30	30	31	31
29	29	30	30	30	30	31	31	31	32
30	30	31	31	31	31	32	32	33	35
31	32	32	32	33	34	35	36	37	38
32	33	34	35	35	36	36	37	38	40
33	35	35	36	36	37	38	38	40	41
34	35	36	36	37	37	38	39	40	42
35	36	36	37	37	38	39	40	41	42
36	37	38	38	39	40	41	41	42	43
37	38	39	39	40	41	42	42	43	43
38	39	40	41	41	42	42	43	43	44
39	40	41	41	42	42	43	43	44	44
40	41	42	42	42	43	43	44	44	45
41	42	42	42	43	43	44	44	45	46
42	42	43	43	43	44	44	45	46	47
43	44	44	44	45	45	46	47	47	48
44	45	45	45	46	47	47	48	48	49
45	46	47	47	47	48	48	49	49	50
46	47	47	47	48	48	49	49	50	50
47	48	48	49	49	49	50	50	51	51
48	49	49	49	50	50	50	51	51	52
49	50	50	50	51	51	51	52	52	52
50	51	51	51	51	52	52	52	53	53
51	51	52	52	52	52	53	53	53	54
52	52	53	53	53	53	53	54	54	54
53	54	54	54	54	54	54	55	55	55
54	54	55	55	55	55	55	56	56	56
55	55	55	55	56	56	56	56	56	57
56	56	56	57	57	57	58	59	59	60
57	58	59	59	60	60	61	62	63	63
58	59	60	60	61	62	62	63	63	64
59	60	61	61	62	62	63	64	64	65
60	61	61	62	63	63	64	64	65	65
61	62	62	63	63	64	65	65	66	66
62	63	63	64	64	65	65	66	66	67
63	64	64	65	65	65	66	66	67	68
64	65	65	65	66	66	66	67	67	68
65	65	66	66	66	67	67	68	68	69
66	67	67	68	68	68	69	69	69	70
67	68	68	68	69	69	69	70	70	70
68	68	69	69	69	70	70	70	71	71
69	70	70	70	70	71	71	71	72	72
70	70	71	71	71	71	72	72	73	73
71	72	72	72	73	73	73	74	74	74
72	73	73	73	73	74	74	74	74	75
73	73	74	74	74	74	75	75	75	76
74	74	75	75	75	75	76	76	76	76
75	75	76	76	76	76	77	77	77	77
76	76	76	77	77	77	77	77	77	77
77	77	77	77	78	78	78	78	78	78
78	78	78	78	79	79	79	79	80	80
79	79	80	80	80	80	.013	.012	.012	.012
80	80	.013	.013	.013	.012	.012	.012	.012	.011

Figure 2-4. Drill sizes for units above sea level

unit efficiency. The standard formula to find the TD is as follows:

$$TD = \frac{(Btuh)(Efficiency)}{(Cooling\ cfm)(1.08)}$$

The Btuh input for electric, natural gas, LP gas, and oil-fired heating units has already been discussed. The efficiencies for each must now be discussed.

Electricity

The output in heat energy is equal to the electric heat energy input. The electric heat element is 100% efficient, so the efficiency multiplier in the formula is 1.

Gas and Oil

To find the efficiency of gas-fired, natural gas or LP gas, or oil heating units, two measurements must be obtained: CO_2 content of the flue products and temperature of the flue products.

For CO_2 content, a CO_2 test must be performed after making sure the input to the unit is within the correct range. The unit also must operate long enough for the thermometers to stabilize while measuring the entering and leaving air temperatures. If the unit has an open diverter on the outlets of the heat exchanger sections, the CO_2 analyzer sampling tube must be inserted far enough into the heat exchanger outlets to eliminate the possibility of outside air mixing with the flue products being tested.

Figure 2-5 shows the use of a CO_2 analyzer on a gas-fired unit. The metal probe is usually 1/4-in. copper tube that is long enough (usually 14 in.) to reach beyond the flue restrictors into the upper portion of the heat exchanger section. For an accurate CO_2 sample, pump the rubber bulb to clear the tube, bulb, and gauge of air. Then invert the gauge several times to force the vapor and liquid to mix. The liquid absorbs the CO_2 from the flue sample, reducing the volume of vapor in the gauge. Held upright, the liquid is drawn up the gauge to the reduced volume. This gives the percentage of CO_2 in the sample.

The CO_2 analyzer is used for natural and LP gas units and oil-fired units. On oil-fired units, the flue sample must be taken from the smoke pipe section between the unit flue outlet and the barometric draft control, as close to the heating unit as possible. This procedure reduces the possibility of fresh air entering the barometric draft control and diluting the flue sample.

To obtain the temperature of flue products, take the stack temperature as shown in Figure 2-6. The location must be as far away as possible from any fresh air that could enter the flue products so that the stack temperature will not be reduced.

Figure 2-7 shows large dial-type thermometers with 12-in. stems for measuring stack temperature. The temperature range is from 200° to 1000°F.

Figure 2-8 shows combustion efficiency slide rules that may be used on standard natural gas units, high-efficiency natural gas units, LP units, fuel oil, and solid fuel (coal) units. These slide rules use the CO_2 and stack temperature information to determine the operating efficiency of the heating unit.

Hand-operated analyzers and mechanical thermometers may be used for combustion testing, but they sacrifice some accuracy. For more accurate results, electronic instruments are recommended. Figure 2-9 shows a portable combustion analyzer that reads and calculates combustion efficiency for six different fuels with direct reading displays. Accurate instruments such as this are a necessity for adjustment of today's high-efficiency heating units.

Figure 2-10 shows the method of measuring the temperature increase through the unit at the return air and supply air duct connections to the unit. Measuring the temperature of the

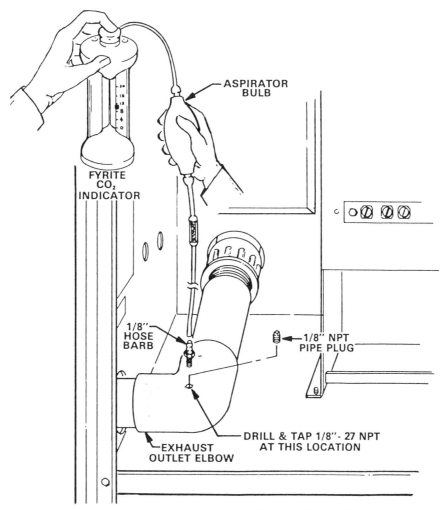

Figure 2-5. Using CO_2 analyzer on a gas-fired unit (Courtesy, Lennox Industries, Inc.)

Figure 2-6. Checking stack temperature (Courtesy, Bacharach Inc.)

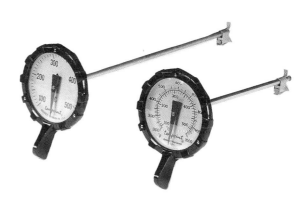

Figure 2-7. Stack temperature thermometers (Courtesy, Bacharach Inc.)

Figure 2-8. Combustion efficiency slide rules (Courtesy, Bacharach Inc.)

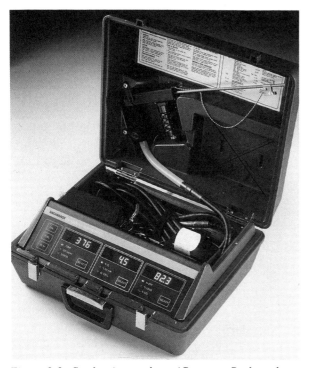

Figure 2-9. Combustion analyzer (Courtesy, Bacharach Inc.)

air at the return air register and supply air grille will not produce accurate results, because air leaks from the ducts. However, the amount of heat gain and/or loss from the duct system can be determined by measuring the temperatures at the return and supply air grilles while measuring the temperatures at the unit. If more than a 2°F difference is found between the unit and either or both of the duct sections, additional insulation is required on the duct after it is completely sealed. If a total of 4°F difference is found in the supply and return ducts in the cooling mode of the air conditioning (a/c) system or heat pump, this represents a 20% loss in system capacity and a 20% increase in operating cost.

CALCULATING THE AIR TEMPERATURE CHANGE

Now that the Btuh input and unit efficiencies for various units have been discussed in detail, it is possible to calculate the difference in temperature (TD) method.

Example 2-2. An 80,000 Btuh natural gas unit is operating at 80% efficiency and 973 cfm. What is the difference in temperature in the unit?

Solution 2-2. Apply the formula given earlier:

$$TD = \frac{(Btuh)(Efficiency)}{(Cooling\ cfm)(1.08)}$$

$$TD = \frac{(80,000\ Btuh)(.80)}{(973\ cfm)(1.08)} = 60.9°F$$

Adjust the blower to obtain a difference in temperature that equals the TD just calculated. This will result in the standard conditions listed earlier in this chapter of 400 cfm per 12,000 Btuh cooling capacity.

Various instruments may be used to determine how much air (cfm) is flowing through a duct distribution system. The results are not as accurate as using the temperature increase method (especially with electric heat), but the temperature increase method is not always available.

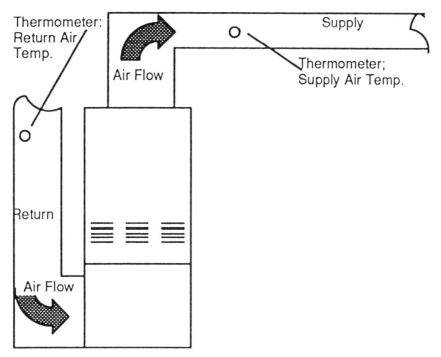

Figure 2-10. Measuring air temperature rise (Courtesy, Inter-City Products Corporation)

DUCT VELOCITY

To establish the desired cfm required by the cooling unit at 95°F outdoor temperature, the velocity of the air flowing through the duct is determined. This is found by dividing the desired cfm by the square foot (ft^2) area of the duct:

$$\text{Duct velocity} = \frac{\text{cfm}}{\text{Duct area (ft}^2\text{)}}$$

Example 2-3. Air at 973 cfm is being supplied through an 8" x 20" main duct. Calculate the duct velocity in feet per minute (fpm) and determine the velocity pressure.

Solution 2-3. It is first necessary to find the area of the duct:

$$\text{ft}^2 = \frac{(\text{height})(\text{width})}{144 \text{ in}^2/1 \text{ ft}^2}$$

$$\text{ft}^2 = \frac{(8")(20")}{144 \text{ in}^2/1 \text{ ft}^2} = 1.11$$

Now it is possible to plug these numbers into the original formula to obtain the duct velocity:

$$\text{Duct velocity (fpm)} = \frac{973 \text{ cfm}}{1.11 \text{ ft}^2} = 876 \text{ fpm}$$

The velocity pressure table in Figure 2-11 shows that an 800 fpm velocity requires a velocity pressure (Vp) of 0.04 in. wc. A velocity of 895 fpm requires a velocity pressure of 0.05 in. wc. The velocity pressure for 876 fpm is not in the velocity pressure table. To find the velocity pressure, use the two closest velocity figures in the table. The difference between the two closest velocity figures is 95 fpm (895 fpm - 800 fpm) and 0.01 in. wc (0.05 in. wc - 0.04 in. wc).

The difference between 876 fpm and 800 fpm is 76 fpm. Determine the percent of 0.01 in. wc that 76 fpm comprises (remember, 95 cfm equals 0.01 in. wc for this example), which means dividing 76 fpm by 95 fpm, which is 0.8. The difference between the velocity

Adjusting Airflow 23

Velocity Pressure (Vp)	Velocity	Velocity Pressure (Vp)	Velocity
.01	400	.31	2230
.02	565	.32	2260
.03	695	.33	2300
.04	800	.34	2335
.05	895	.35	2370
.06	980	.36	2400
.07	1060	.37	2435
.08	1130	.38	2470
.09	1200	.39	2500
.10	1265	.40	2530
.11	1325	.41	2560
.12	1385	.42	2595
.13	1440	.43	2620
.14	1500	.44	2655
.15	1550	.45	2680
.16	1600	.46	2715
.17	1650	.47	2740
.18	1700	.48	2775
.19	1745	.49	2800
.20	1790	.50	2830
.21	1835	.51	2860
.22	1880	.52	2880
.23	1920	.53	2915
.24	1960	.54	2940
.25	2000	.55	2970
.26	2040	.56	2995
.27	2080	.57	3020
.28	2120	.58	3050
.29	2155	.59	3075
.30	2190	.60	3100
.61	3125	.81	3600
.62	3150	.82	3620
.63	3180	.83	3640
.64	3200	.84	3600
.65	3225	.85	3685
.66	3250	.86	3710
.67	3280	.87	3730
.68	3300	.88	3760
.69	3325	.89	3775
.70	3350	.90	3800
.71	3370	.91	3820
.72	3395	.92	3840
.73	3410	.93	3860
.74	3440	.94	3880
.75	3465	.95	3900
.76	3490	.96	3920
.77	3510	.97	3940
.78	3540	.98	3960
.79	3560	.99	3980
.80	3580	1.00	4000

Note: Table is based on standard air (dry air at 70°F and 29.92 inches of mercury absolute barometric pressure)

Figure 2-11. Velocity pressures

pressure for 800 fpm and 876 fpm is 80% of the difference between the in. wc for 800 fpm and the in. wc for 876 fpm. Eighty percent of 0.01 in. wc is 0.008 in. wc (0.01 in. wc x 0.80). Add 0.008 in. wc to the 0.04 in. wc for 800 fpm to obtain a velocity pressure of 0.048 in. wc. Therefore, a velocity pressure of 0.048 in. wc will be needed to push the desired cfm through the duct.

MANOMETERS

Figure 2-12 shows an *inclined manometer* for measuring air pressure in a range of -.10 in. wc to +1.0 in. wc. The top has a hose tap on the left for positive pressure that measures the pressure in the supply portion of the duct system. This pressure forces the liquid down the inclined column. The top also has a hose tap on the right that measures negative pressure in the return side of the duct system. This negative pressure draws the liquid into the inclined gauge.

When only the top left tap is used, this instrument measures the pressure in the supply duct compared to the atmospheric pressure surrounding the duct. Using only the top right hose tap, the instrument measures the pressure in the return duct compared to the atmospheric pressure surrounding the duct. This is a pressure below atmospheric pressure — a negative pressure. When using both taps, the instrument measures the total pressure in the duct system. This is called the *external static pressure* (ESP) of the unit.

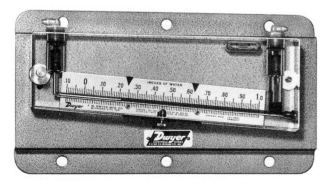

Figure 2-12. Inclined manometer (Courtesy, Dwyer Instruments, Inc.)

ESP will be discussed in more detail later in this chapter.

The inclined manometer is used with a *pitot tube* to measure the velocity pressure of the air through a duct. A pitot tube is a double tube with two hose connections to the manometer. Figure 2-13 shows a method of connecting the pitot tube to the manometer. The inside tube measures the pressure of the air moving against the end of it. This pressure is made up of the air pressure in the duct and the forward movement of the air traveling through the duct. This is called the *total air pressure* (Pt) in the duct. The side openings on the outside tube measure the *static pressure* (Ps) in the duct. The difference between the total pressure and the static pressure is the velocity pressure of the air moving through the duct.

Connecting the pitot tube so that the total pressure forces the liquid down the manometer scale and so that the static pressure forces the liquid up the scale makes the manometer indicate the difference between the two pressures — the velocity pressure of the air moving through the duct.

When using a pitot tube to measure duct velocity, it is necessary to take several readings across the duct, because air in different places does not travel at the same velocity. Air close to the surface of the duct travels at a slower speed than the air in the middle of the duct. Taking a series of readings is called *traversing* the duct. The duct area is divided into sections of equal size and the velocity pressure is measured in each section. Figure 2-14 shows a rectangular 20" x 10" duct divided into eight sections. A reading should be taken from each of the sections, and an average velocity pressure can then be calculated.

In a round duct, measurements are taken from the center of the duct to the perimeter in a minimum of eight directions. Figure 2-15 shows the positions that would be used in a 14-in. diameter round duct. In either round or rectangular ducts, the larger the duct, the

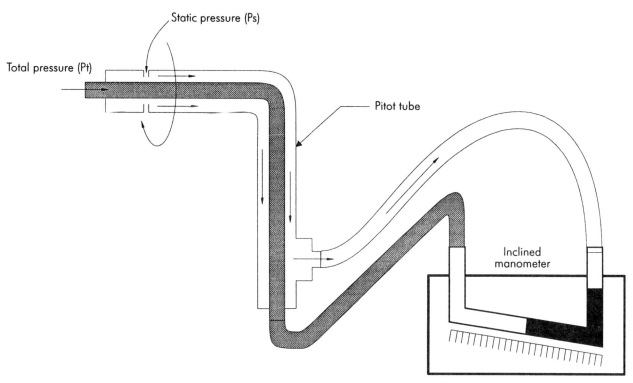

Figure 2-13. Pitot tube and manometer measuring velocity pressure

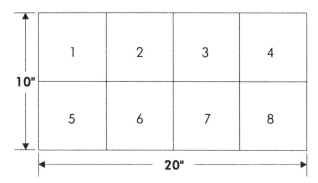

Figure 2-14. Rectangular duct traverse pattern for measuring duct velocity

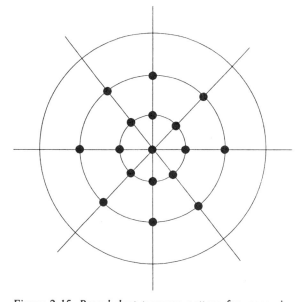

Figure 2-15. Round duct traverse pattern for measuring duct velocity

more readings that must be obtained for the highest accuracy in the calculations.

Pressure readings must be taken in straight sections of the duct downstream from any bends or restrictions. The distance between any bend or restriction and the pressure tap must be a minimum of 1.5 times the greater dimension of the duct. If, for example, the duct is an 8" x 20" duct, the minimum distance downstream from a turn in the duct would be 30 in.

Example 2-3 showed how to determine the area of a rectangular duct and the subsequent velocity. To determine the area of a round duct, use the following formula:

Area = πr^2

where: Area = cross sectional area of the duct

π = ratio between the circumference of a round circle and the diameter of the circle

This ratio is 3.416; therefore, the circumference of a circle is 3.416 times the diameter.

r = one-half the diameter of a circle

The area of a 14-in. round duct is calculated as follows:

$$\text{Area} = \frac{(3.416)(7")(7")}{144 \text{ in}^2/1 \text{ ft}^2} = 1.16 \text{ ft}^2$$

For the round duct, using the 973 cfm, the calculation of the velocity would be:

$$\text{Duct velocity (fpm)} = \frac{973 \text{ cfm}}{1.16 \text{ ft}^2} = 839 \text{ fpm}$$

Using Figure 2-11, find the velocity pressure (see the explanation on how to calculate velocity pressure in Example 2-3). Calculations show that 839 fpm air velocity through the duct results in a desired average velocity pressure through the duct of 0.0441 in. wc.

EXTERNAL STATIC PRESSURE (ESP)

Most units are designed to supply air at a total ESP of .25 in. wc without a/c added. If an a/c or heat pump evaporator is added into the air distribution system, the unit must be able to supply the required amount of air against an ESP of 0.50 in. wc, which is the industry design standard. The manufacturer usually includes cfm tables based on the ESP against the unit.

Figure 2-16 illustrates the cfm for various heat pump models that operate under total external static pressures ranging from 0.10 to 0.50 in. wc. The total external static pressure is made up of both the positive supply pressure and the negative return pressure totaled as though each number is a positive number. For example, a supply pressure of +0.20 in. wc and a return pressure of -0.05 in. wc totals +0.25 in. wc.

Again, these pressures must be taken in straight sections of the duct downstream from any bends or restrictions. The distance between any bend or restriction and the pressure tap must be a minimum of 1.5 times the greater dimension of the duct. This type of information should be gathered and retained by the service technician to help in the fine tuning process.

When the manufacturer's information is available, this information should be used to determine the cfm through the indoor air handling section. To make a comparison of the unit capacity to the manufacturer's specifications, the unit must have 400 cfm per 12,000 Btuh per the unit rated capacity of 12,000 Btuh. If the unit is rated at 36,000 Btuh, the required cfm would be 400 cfm per 12,000 Btuh, or 1200 cfm. By adjusting the ESP of the AH65HF indoor section shown in Figure 2-16 to 0.21 in. wc, it is possible to achieve 1200 cfm through the indoor section.

Remember, the total ESP is composed of the positive pressure in the supply duct plus the negative pressure in the return duct. Controlling either one to obtain the desired ESP will give the desired cfm for the comparison test.

TOTAL CFM MEASUREMENT

If the temperature difference method is not available, another method that may be used to determine the total cfm through the evaporator is to measure the cfm from each

INDOOR SECTION SPECIFICATIONS

MODEL	BLOWER MOTOR	EVAPORATOR COIL CFM (1) @ EXTERNAL STATIC PRESSURES (INS. OF WATER)							WEIGHT	
		.10	.15	.20	.25	.30	.40	.50	NET	SHIP
AH65HD	¼	870	840	810	780	750	690	620	113	128
AH65HE	½	1260	1240	1220	1195	1170	1100	1025	121	136
AH68HF	½	1410	1380	1350	1310	1290	1220	1150	158	173
AH68HG	¾	1760	1740	1720	1685	1650	1575	1450	155	170
AH68HH	¾	1870	1845	1830	1790	1750	1690	1590	160	175
AH68HK	¾	1870	1845	1830	1790	1750	1690	1590	170	186

(1) Since filter furnished with unit, total E.S.P. available to ducts and grills.

Figure 2-16. Indoor air handler specifications (Courtesy, Addison Products Company)

supply grille in the supply duct system. There are many types of instruments used to measure air velocities and calculate cfm from supply or return grilles.

ANEMOMETERS

Anemometers, which measure air speed, are available in many types and arrangements and can either be mechanical and digital.

Figure 2-17 shows a mechanical vane-type anemometer used to measure the velocity of air from supply or return outlets. When held over the face of a grille for a total of one minute, the instrument indicates the velocity of the air. The entire face of the supply or return grille must be tested to obtain accurate results. By pausing with the instrument over each portion of the grille for an equal period of time, the reading for the one-minute period will be the average velocity of the air through the grille. This movement is necessary, because the air velocity is not always equal in different sections of the grille. If the grille is at the end of a straight duct, the velocity of the air through the various parts of the grille will be closer to equal compared to air flowing through a grille at the end of a duct that has a bend. For example, the air distributed through a grille located at the outlet of a stackhead fitting will have 75% of the air out of the upper 50% of the grille's free area (the total opening between grille bars).

Figure 2-17. Vane anemometer (Courtesy, Airflow Technical Products, Inc.)

The average velocity of the air in fpm multiplied by the free area of the grille in ft^2 will equal the cfm of air flowing through the grille. Refer to manufacturer specifications for the free area of the grille.

Figure 2-18 shows specifications for one manufacturer's sidewall registers, grilles, and ceiling outlets. Each size shows the free area in square inches. Use these figures to find the air in cfm coming from the grille. To do this, convert the free area figure in square inches to square feet (divide by 144 in²/1 ft²). Then multiply that figure by the average velocity of the air from the grille. For example, according to Figure 2-18, a 14" x 6" register has 62 in² free area. Convert the free area in in² to ft².

$$\frac{62 \text{ in}^2}{144 \text{ in}^2/\text{ft}^2} = .4305 \text{ ft}^2$$

The measured average velocity for this register is 290 fpm, so the cfm from the register is as follows:

$(.4305 \text{ ft}^2)(290 \text{ fpm}) = 125 \text{ cfm}$

Figure 2-19 shows an electronic or digital-type anemometer that measures the velocity of the air with an electronic sensing probe. This device measures air velocity by means of the heat loss from a heat source in the probe. The loss is measured and converted in the solid state circuitry to current flow through the meter that registers in fpm. This instrument must also be swept across the grille face to obtain the average velocity reading. Calculation is based on the grille's free area in square feet.

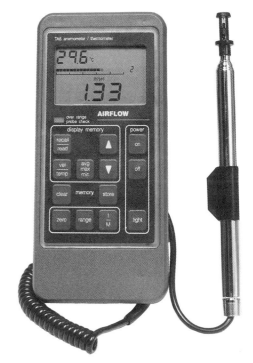

Figure 2-19. Digital-type anemometer (Courtesy, Airflow Technical Products, Inc.)

Flowhoods

The easiest and most accurate way to measure cfm is by using a flowhood such as the one shown in Figure 2-20. The flowhood reads the cfm by directing all the air from the grille through its gathering hood. An

series 10L, 10M, 10BIL

- Series 10B1L not available in 8" x 6"
- For overall O.D. on 10L & 10M add 1½" to nominal listed size

Product Number and Size	FREE AREA SQ. IN.	Heating BTU/h	3045	4565	6090	7610	9515	11415	13320	15220	17125	19025	20930	22830	24735
		Cooling BTU/h	855	1280	1710	2135	2670	3200	3735	4270	4805	5340	5870	6405	6940
		C.F.M.	40	60	80	100	125	150	175	200	225	250	275	300	325
10L,10M – 8" x 6"	34	T.P. Loss	.007	.013	.021	.031	.050	.078	.101						
		Horiz. Throw (ft.)	3.5	5	6.5	8	11	15	19						
		Velocity	170	255	340	425	530	635	745						
10L,10M,10B1L – 10" x 6"	43	T.P. Loss		.007	.012	.020	.028	.041	.056	.083	.109				
		Horiz. Throw (ft.)		4.5	6	7.5	10	12	15	18.5	21.5				
		Velocity		200	265	335	415	500	585	665	750				
10L,10M,10B1L – 12" x 6"	53	T.P. Loss			.007	.012	.021	.029	.042	.060	.081	.095	.107		
		Horiz. Throw			4.5	6	8	10	12	14	16.5	20	23.5		
		Velocity			220	275	340	410	480	560	615	680	750		
10L,10M,10B1L – 14" x 6"	62	T.P. Loss				.008	.015	.023	.033	.041	.053	.077	.092	.103	.119
		Horiz. Throw				5	6	9	10.5	12	14	16.5	18.5	21.5	24.5
		Velocity				230	290	350	410	465	525	580	640	695	755

Figure 2-18. Supply register specifications (Courtesy, Milcor/Lima Register)

Figure 2-20. Flowhood (Courtesy, Shortridge Instruments Inc.)

opening in the base of the instrument contains a meter, which is calibrated to read air quantity in cfm.

The flowhood is made in several sizes and configurations. In residential and small commercial testing, the 2' x 2' size is the most popular. The cfm measurement range is from 25 to 2599 cfm of supply air and 25 to 1500 cfm of return air. The airflow in total cfm is obtained by measuring the cfm at each supply grille.

PROBLEMS

2.1. What is the significance of 1.08?

2.2. What is the total heat output of a 4.0 kWh electric heating element?

2.3. What is the heat output of an 80% efficient gas-fired unit rated at 120,000 Btuh?

2.4. What is the heat output of an 80% efficient oil-fired unit using a 0.85 gph nozzle?

2.5. The utility supplies natural gas with 1050 Btu/ft^3. The meter's 1-ft^2 dial revolves once every 39 seconds. What is the output of the 80% efficient heating unit?

2.6. A gas-fired heating unit with a manufacturer's rating of 80,000 Btuh is installed at 8000-ft altitude. What is the proper input rate for this unit?

2.7. An oil-fired heating unit that is rated at 119,000 Btuh will require what nozzle size?

2.8. The average velocity of the air flowing through an 18-in. round duct is 860 fpm. How much air in cfm is flowing through the duct?

2.9. The average velocity of air flowing through a 6" x 14" supply grille with 65% free area is 480 fpm. How many cfm are being supplied through the grille?

2.10. The input tolerance for fossil fuel-powered heating units is:

    A. -10% to +10%

    B. -10% to +5%

    C. -5% to +5%

    D. -0% to +10%

    E. -10% to +0%

CHAPTER THREE

Determining Refrigerant Charge

Chapter 2 discussed the different methods used to obtain the correct airflow in cfm for various units. No matter which method is used, the unit should operate according to manufacturer specifications. To produce the desired heat transfer capacity, not only must the proper amount of air pass through the heat exchangers, the refrigerant charge in the system must be correct.

After obtaining the desired airflow in cfm through the evaporator, system pressures and temperatures will adjust to reflect the refrigerant charge. If a capillary tube is used, the amount of refrigerant charge will be revealed by the superheat of the vapor from the evaporator or by the subcooling of the liquid refrigerant from the condenser coil. If the system uses thermostatic expansion valves (TXVs), the refrigerant quantity will only be reflected in the amount of subcooling from the condenser coil.

To determine the correct amount of refrigerant charge, it is necessary to be familiar with all aspects of the system. Always read the manufacturer literature, which contains the compressor operating pressures, as well as the superheat for the cooling mode; liquid subcooling for the cooling mode; net capacity for the cooling mode; and gross capacity for the heating mode.

In the cooling mode, the operating suction and discharge pressures are determined by the dry bulb temperature of the air entering the outdoor unit condenser coil and the wet bulb temperature of the air entering the indoor evaporator in the a/c unit or heat pump. In the heating mode, the suction and discharge pressures in a heat pump are determined by the dry bulb temperature of the air entering the indoor condenser coil and the dry bulb temperature of the air entering the outdoor evaporator.

Figure 3-1 is a page from an installation, operation, and maintenance manual covering a 2-horsepower (hp) split system heat pump. A variety of information is given for this unit in the cooling mode, including the following:

* The system is rated at 800 cfm through the evaporator (400 cfm per 12,000 Btuh cooling capacity).

* Performance data is given for an outdoor dry bulb temperature range from 65° to 115°F in combinations with indoor air enthalpy from 63° to 69°F wet bulb temperatures.

* Included at the bottom of the table are factors by which the amps in the tables can be multiplied to correct for voltage variations at the unit.

- Desired difference in temperature across the evaporator (DELTA T EVAP in the chart) is included.

- The chart lists rated Btuh capacity. These tables are limited to the system using capillary tubes or restrictors. Systems using TXVs do not have the wide suction pressure swings from variations in condenser inlet air dry bulb temperatures and variations in evaporator inlet wet bulb temperatures (enthalpy).

- Performance in the heating mode is also included. With no humidity involved in the performance, the variations are only due to changes in the dry bulb temperature of the air entering the outdoor evaporator.

Presenting the information in this manner shows that if the suction and discharge pressures are correct for the given outdoor and indoor conditions, the system should produce the rated capacity. Always use manufacturer information when it is available.

SUPERHEAT

Determining superheat is one method of ensuring the system contains the proper refrigerant charge. Superheat always refers to a vapor. A superheated vapor is any vapor that is above its saturation temperature for a given pressure. All liquid must be vaporized for superheating to occur; any additional heat will cause the 100% saturated vapor to start superheating. This additional heat will cause the vapor to increase in temperature and gain sensible heat. Superheating is a sensible heat process. Superheated vapor occurs in the evaporator, suction line, and compressor.

The amount of superheat that builds up in the evaporator is dependent upon the quantity of refrigerant forced into the evaporator to absorb heat and the amount of air over the evaporator to supply the heat. With the proper quantities of air and refrigerant, the pressures in the system will adjust to provide optimum heat absorption and superheat.

The flow of refrigerant into the evaporator will depend upon the pressure in the condenser coil. This pressure forces the liquid refrigerant into the capillary tube(s) or restrictor and is opposed by the pressure in the evaporator that results from the expanding refrigerant absorbing heat in the evaporator.

The pressure in the condenser coil, the *condensing temperature*, is determined by how much vapor, or heat, is received from the compressor compared to the temperature of the air entering the condenser coil and extracting the heat. The amount of vapor from the compressor will depend upon the amount of refrigerant vaporized in the evaporator. This will depend upon how much air in cfm is going through the evaporator and the total heat content of the air, both sensible and latent heat.

The amount in cfm has been established by proper adjustment of air quantity to obtain the desired cfm for the particular system either by the 400 cfm per 12,000 Btuh or the manufacturer's specifications. Therefore, the load on the evaporator and the outdoor air (ambient) temperature will determine the amount of superheat developed in the evaporator. To be more specific, the amount of superheat is determined by the outdoor temperature and the wet bulb temperature of the air entering the evaporator.

Superheat is measured as follows:

- *Compressor suction pressure.* The suction pressure will give the boiling point of the refrigerant at the point where the pressure is measured.

- *Refrigerant boiling point at the evaporator.* When the suction pressure is measured at the outdoor section of the system or at the compressor line tap in a package unit, the flow resistance of

COOLING   INDOOR CFM 800

WET BULB INDOOR	OUTDOOR 65 DEGREES				OUTDOOR 75 DEGREES			
	63	65	67	69	63	65	67	69
SUCT PSIG	55.6	56.3	57.4	58.2	62.0	62.8	63.9	64.8
LIQ PSIG	151	154	156	159	186	189	192	196
AMPS*	9.0	9.0	9.2	9.5	10.0	10.1	10.2	10.6
DELTA T EVAP	20.4	19.2	18.0	16.5	21.4	20.2	19.0	17.4
BTUH	22,669	23,146	23,862	24,816	23,604	24,101	24,846	25,840
S/T RATIO	0.78	0.72	0.65	0.57	0.78	0.73	0.66	0.58
SENSIBLE x 1000	17.6	16.6	15.6	14.3	18.5	17.5	16.4	15.0

WET BULB INDOOR	OUTDOOR 85 DEGREES				OUTDOOR 95 DEGREES			
	63	65	67	69	63	65	67	69
SUCT PSIG	68.9	69.8	71.1	72.1	75.2	76.1	77.5	78.6
LIQ PSIG	210	213	216	221	233	236	240	245
AMPS*	10.8	10.9	11.1	11.5	11.5	11.6	11.8	12.3
DELTA T EVAP	22.4	21.1	19.8	18.1	21.9	20.7	19.4	17.7
BTUH	23,305	24,816	25,584	26,607	23,370	23,862	24,600	25,584
S/T RATIO	0.79	0.73	0.67	0.59	0.81	0.75	0.68	0.60
SENSIBLE x 1000	19.3	18.2	17.1	15.6	18.9	17.8	16.7	15.3

WET BULB INDOOR	OUTDOOR 105 DEGREES				OUTDOOR 115 DEGREES			
	63	65	67	69	63	65	67	69
SUCT PSIG	77.9	78.8	80.3	81.4	78.4	79.4	80.8	82.0
LIQ PSIG	263	267	271	277	286	291	295	302
AMPS*	12.1	12.2	12.4	12.9	12.5	12.7	12.9	13.4
DELTA T EVAP	20.8	19.7	18.4	16.9	19.7	18.6	17.4	15.9
BTUH	20,799	21,237	21,894	22,770	18,696	19,090	19,680	20,467
S/T RATIO	0.87	0.80	0.73	0.64	0.91	0.84	0.77	0.67
SENSIBLE x 1000	18.0	17.0	15.9	14.6	17.0	16.1	15.1	13.8

HEATING
ALL DATA IS FOR 70° INDOOR WITH ICE-FREE OUTDOOR COIL

AMBIENT - °F	-10	0	10	20	30	40	50	60	70
HEAT BTUH	11832	12064	12528	12992	14152	17400	25984	32480	35728
UNIT COP	1.89	1.92	1.98	2.01	2.10	2.43	3.24	4.02	4.35
DELTA T - °F	15.9	16.2	16.8	17.5	19.0	23.4	34.9	43.6	48.0
DISCH. PSIG @ UNIT	116	133	148	162	173	183	191	198	201
SUCT. PSIG @ COMPR.	12.3	17.0	22.9	29.9	38.0	47.2	57.5	68.9	81.5
OUTDOOR AMPS	6.2	6.7	7.1	7.5	8.0	8.6	9.5	10.7	12.4

*TO CORRECT FOR VOLTAGE VARIATIONS AT THE UNIT, MULTIPLY THE AMPS IN THE TABLE BY THE FOLLOWING CORRECTION FACTOR:

   1 PHASE: 208V = 1.10, 220V = 1.05, 230V = 1.00, 240V = .95

ALL VALUES ARE FOR SERVICE ONLY AND MAY VARY UP OR DOWN FROM TABLE VALUES.

Figure 3-1. Unit performance table (Courtesy, Addison Products Company)

the vapor through the line must be considered. Suction lines should be sized to have a pressure of 3 psig per 100 equivalent feet of line. A fairly safe assumption is 2 psig per 100 equivalent feet for the average size installation. By adding 2 psig to the suction pressure reading, the operating pressure in the evaporator is found. This pressure, when converted to temperature on the temperature-pressure chart, Figure 3-2, for the particular refrigerant, will give the boiling point. For example, if the system uses R-22 and the measured suction pressure is 82 psig, adding 2 psig pressure loss in the suction line will mean there is an 84 psig operating pressure in the evaporator. Using the temperature-pressure chart in Figure 3-2, the saturation temperature, or boiling point, of the R-22 refrigerant is 50°F.

- *Suction line temperature at the evaporator outlet.* After the liquid refrigerant vaporizes, the vapor temperature will continue to rise depending upon how long it has to travel from the place where the last of the liquid refrigerant evaporates, the point of vaporization, to the evaporator outlet. The physical temperature of the refrigerant at this point will reflect the heat gain.

- *Evaporator superheat.* The amount of superheat is the difference in temperature between the physical temperature of the refrigerant and the evaporating temperature, or boiling point, measured in the previous two points. If the physical temperature of the vapor from the evaporator is 61°F with a 50°F refrigerant boiling point, the superheat will be 61°F less 50°F, or 11°F.

- *Temperature of the air entering the condenser coil.* The flow rate of the refrigerant depends upon the head pressure. The head pressure is affected by the outdoor dry bulb temperature, or the condenser air inlet temperature. The outdoor dry bulb temperature must be determined at the same time the other readings are taken. This temperature can change in short time periods, which is why a reading must be taken for each test.

- *Return air wet bulb temperature.* The wet bulb temperature of the return air supply is determined when adjusting the

SPORLAN TEMPERATURE PRESSURE CHART

Vacuum-Inches of Mercury – Italic Figures
Pressure-Pounds Per Square Inch Bold Figures

TEMPERATURE °F.	12-F	22-V	500-D	502-R	717-A	TEMPERATURE °F.	12-F	22-V	500-D	502-R	717-A	TEMPERATURE °F.	12-F	22-V	500-D	502-R	717-A
-60	19.0	12.0	17.0	7.2	18.6	12	15.8	34.7	21.2	43.2	25.6	42	38.8	71.5	48.2	83.8	61.6
-55	17.3	9.2	15.0	3.9	16.6	13	16.5	35.7	21.9	44.3	26.5	43	39.8	73.0	49.4	85.4	63.1
-50	15.4	6.2	12.8	0.2	14.3	14	17.1	36.7	22.6	45.4	27.5	44	40.7	74.5	50.5	87.0	64.7
-45	13.3	2.7	10.4	1.9	11.7	15	17.7	37.7	23.4	46.5	28.4	45	41.7	76.0	51.6	88.7	66.3
-40	11.0	0.5	7.6	4.1	8.7	16	18.4	38.7	24.1	47.7	29.4	46	42.7	77.6	52.8	90.4	67.9
-35	8.4	2.6	4.6	6.5	5.4	17	19.0	39.8	24.9	48.9	30.4	47	43.6	79.2	54.0	92.1	69.5
-30	5.5	4.9	1.2	9.2	1.6	18	19.7	40.9	25.7	50.0	31.4	48	44.7	80.8	55.1	93.9	71.1
-25	2.3	7.4	1.2	12.1	1.3	19	20.4	41.9	26.5	51.2	32.5	49	45.7	82.4	56.3	95.6	72.8
-20	0.6	10.1	3.2	15.3	3.6	20	21.0	43.0	27.3	52.5	33.5	50	46.7	84.0	57.6	97.4	74.5
-18	1.3	11.3	4.1	16.7	4.6	21	21.7	44.1	28.1	53.7	34.6	55	52.0	92.6	63.9	106.6	83.4
-16	2.1	12.5	5.0	18.1	5.6	22	22.4	45.3	28.9	54.9	35.7	60	57.7	101.6	70.6	116.4	92.9
-14	2.8	13.8	5.9	19.5	6.7	23	23.2	46.4	29.8	56.2	36.8	65	63.8	111.2	77.8	126.7	103.1
-12	3.7	15.1	6.8	21.0	7.9	24	23.9	47.6	30.6	57.5	37.9	70	70.2	121.4	85.4	137.6	114.1
-10	4.5	16.5	7.8	22.6	9.0	25	24.6	48.8	31.5	58.8	39.0	75	77.0	132.2	93.5	149.1	125.8
-8	5.4	17.9	8.8	24.2	10.3	26	25.4	49.9	32.4	60.1	40.2	80	84.2	143.6	102.0	161.2	138.3
-6	6.3	19.3	9.9	25.8	11.6	27	26.1	51.2	33.3	61.5	41.4	85	91.8	155.7	111.0	174.0	151.7
-4	7.2	20.8	11.0	27.5	12.9	28	26.9	52.4	34.2	62.8	42.6	90	99.8	168.4	120.6	187.4	165.9
-2	8.2	22.4	12.1	29.3	14.3	29	27.7	53.6	35.1	64.2	43.8	95	108.3	181.8	130.6	201.4	181.1
0	9.2	24.0	13.3	31.1	15.7	30	28.5	54.9	36.0	65.6	45.0	100	117.2	195.9	141.2	216.2	197.2
1	9.7	24.8	13.9	32.0	16.5	31	29.3	56.2	36.9	67.0	46.3	105	126.6	210.8	152.4	231.7	214.2
2	10.2	25.6	14.5	32.9	17.2	32	30.1	57.5	37.9	68.4	47.6	110	136.4	226.4	164.1	247.9	232.3
3	10.7	26.5	15.1	33.9	18.0	33	30.9	58.8	38.9	69.9	48.9	115	146.8	242.7	176.5	264.9	251.5
4	11.2	27.3	15.7	34.9	18.8	34	31.7	60.1	39.9	71.3	50.2	120	157.7	259.9	189.4	282.7	271.7
5	11.8	28.2	16.4	35.9	19.6	35	32.6	61.5	40.9	72.8	51.6	125	169.1	277.9	203.0	301.4	293.1
6	12.3	29.1	17.0	36.9	20.4	36	33.4	62.8	41.9	74.3	52.9	130	181.0	296.8	217.2	320.8	—
7	12.9	30.0	17.7	37.9	21.2	37	34.3	64.2	42.9	75.9	54.3	135	193.5	316.6	232.1	341.2	—
8	13.5	30.9	18.4	38.9	22.1	38	35.2	65.6	43.9	77.4	55.7	140	206.6	337.3	247.7	362.6	—
9	14.1	31.8	19.0	39.9	22.9	39	36.1	67.1	45.0	79.0	57.2	145	220.3	358.9	264.0	385.0	—
10	14.6	32.8	19.7	41.0	23.8	40	37.0	68.5	46.1	80.5	58.6	150	234.6	381.5	281.1	408.4	—
11	15.2	33.7	20.4	42.1	24.7	41	37.9	70.0	47.1	82.1	60.1	155	249.5	405.1	298.9	432.9	—

Figure 3-2. Temperature-pressure chart (Courtesy, Sporlan Valve Company)

amount of air through the evaporator. This test must be repeated at this time to obtain accurate results. The effect of air temperature changes must be eliminated. Therefore, the wet bulb temperature of the return air must be determined when the outdoor dry bulb temperature, evaporator operating pressure, and the refrigerant vapor temperature are measured.

Once the inside air wet bulb temperature and the outdoor air dry bulb temperature are determined, it is possible to use Figure 3-3 to determine the correct superheat for a capillary tube or restrictor-type system. For example, if the return air wet bulb temperature is 66°F and the outdoor dry bulb temperature is 95°F, the superheat should be 11°F. If the inside air wet bulb temperature is 70°F at the 95°F outdoor dry bulb temperature, the superheat should be 16°F. If the inside air wet bulb temperature is 64°F at the same outdoor dry bulb temperature of 95°F, the superheat should be 7°F.

The greater the total heat content of the air over the evaporator, the faster the refrigerant will boil in the evaporator and the higher the superheat will be. After determining the actual superheat the system is producing and the theoretical superheat desired at the indoor wet bulb and outdoor dry bulb temperatures, a comparison of the two is made. Using average service gauges and thermometers, if the two results are within +/-1°F, the refrigerant charge is correct. The use of digital thermometers is recommended for greater accuracy, such as when the charge must be +/-0.5°F. If the actual superheat is more than the correct superheat, the system is short of refrigerant charge. The addition should be in 4 oz increments, allowing time for the system to reach thermal balance with each increase in refrigerant charge. This will usually occur within a 5 to 15 minute period.

This procedure should be repeated until the superheat is lowered to the correct amount. If the actual superheat is less than the correct superheat, the system is overcharged. The refrigerant should be removed from the liquid line in 4 oz increments, allowing sufficient running time between each refrigerant removal to allow the system to reach thermal balance. Remove the excess refrigerant into a DOT-approved cylinder.

Condenser Air Inlet Temperature °F-dry bulb	Evaporator Air Inlet Temperature °F-wet bulb											
	54	56	58	60	62	64	66	68	70	72	74	76
60	13	17	18	20	24	26	28	30				
65	11	13	17	17	18	22	25	28	30			
70	8	11	12	14	16	18	22	25	28	30		
75		7	10	12	14	16	18	23	26	28	30	
80			6	8	12	14	16	18	23	27	28	30
85				6	8	12	14	17	20	25	27	28
90					6	9	12	15	18	22	25	28
95						7	11	13	16	20	23	27
100							8	11	14	18	20	25
105							6	8	12	15	19	24
110								7	11	14	18	23
115									8	13	16	21

Figure 3-3. Superheat table

Charging or recovery as necessary should be continued until the proper superheat is obtained. Remember to repeat the pressure and temperature tests when the thermal balance has been reached. Changes will occur with the change in refrigerant quantity and temperature.

SUBCOOLING

The amount of subcooling in the condenser coil is another method of checking the refrigerant charge. Subcooling is the removal of heat from a liquid that is below its condensation point. Liquid subcooling occurs when sensible heat is removed from the 100% saturated liquid point in the condenser. Liquid subcooling may occur from the start of the 100% saturated liquid point in the condenser to the metering device. When all of the saturated vapor in the condenser changes to saturated liquid, subcooling will begin if any heat is removed. This method applies to systems using TXVs, capillary tubes, or restrictor-type flow valves, as long as the system does not contain a receiver.

The condenser has a fixed amount of heat exchanger surface to transfer heat from the vapor to the air over the condenser coil surfaces. The objective is to use each portion for the maximum heat extraction in each of the phases of heat transfer. The function of the condenser coil is to:

* lower the temperature of the superheated vapor from the compressor to condensing temperature;

* condense the vapor to a liquid;

* reduce the temperature of the liquid below the condensing temperature.

To realize the importance of subcooling, the entire system performance must be considered. Liquid refrigerant is forced through the metering device to lower its boiling point below the temperature of the air through the evaporator to obtain the desired heat-transfer rate. The vapor is then compressed by the motor-compressor assembly to a pressure and corresponding condensing temperature high enough above the outdoor dry bulb temperature to eject the total amount of heat into the air at the desired rate.

The condenser coil then condenses the vapor into high temperature liquid. This high temperature liquid flows through the liquid line to the metering device, where the liquid pressure and corresponding boiling point are reduced. The physical temperature of the liquid cannot remain higher than the corresponding boiling point. Some liquid refrigerant is vaporized to absorb the heat from the liquid and lower its temperature to the boiling point temperature in the evaporator. The vapor produced in this process is called *flash gas* and represents a reduction in system efficiency. The compressor must handle this vapor along with the vapor produced in the evaporator. To reduce the amount of flash gas produced, the liquid refrigerant temperature is lowered before it leaves the condenser. This is accomplished by having enough liquid refrigerant in the system to have liquid in the last portion of the condenser coil when the system is operating.

If no subcooling took place in the condenser coil, the entire surface would be used for desuperheating (cooling the superheated vapors that the compressor has compressed) and condensing the vapor to liquid. This would produce the heat transfer with a low difference in temperature between the air and the refrigerant vapor. The temperature difference between the condensing temperature and the ambient temperature is called the *condenser split*. On the other hand, the higher temperature of the liquid would result in an abnormally high quantity of flash gas. This would reduce the system capacity by causing the compressor to handle less vapor formed in the evaporator.

Using this information it would seem that the more subcooling that takes place, the higher the system efficiency. This is correct to a point. As refrigerant is added into the system, more condenser coil surface is used for subcooling, which reduces the amount of surface for the vapor-condensing function. As this amount of surface is reduced, the remaining surface must have a higher difference in temperature (split) between the vapor and the air to obtain the heat transfer that is needed. To accomplish this, the condensing temperature is raised, which means the compressor discharge pressure is raised to increase the condensing temperature. As the discharge pressure is raised, the efficiency of the compressor decreases. The reduction in liquid temperature increases the system efficiency while the increase in compressor discharge pressure decreases the system efficiency. It becomes a matter of how much of a change in either one or both of the factors will affect the results.

As subcooling starts to occur, the decrease in liquid temperature (decrease in flash gas) has a greater influence on raising the system efficiency than the detrimental effect of the increase in compressor discharge pressure. As the liquid temperature is reduced, the amount of subcooling is increased. However, the beneficial effect on the efficiency is a diminishing factor. When the compressor discharge pressure is increased, the detrimental effect of this change is an increasing factor. The point at which each effect balances the other is the desirable refrigerant charge. At this point, the net cooling capacity of the system is at its peak.

The amount of refrigerant is the main factor in determining the amount of subcooling at the condenser coil. However, the system must also be operating under the following parameters:

- Outdoor dry bulb temperature range limits of 65° to 115°F

- Inside dry bulb temperatures of 70° to 85°F

- Relative humidity between 35% and 75%

To determine the amount of subcooling, the following test readings need to be taken:

- *Compressor discharge pressure.* The compressor discharge can be measured at the discharge outlet of the compressor if a gauge connection is provided. This will give the pressure that the compressor is producing, although this may not be the actual pressure (condensing temperature) in the condenser coil. On package units, as well as outdoor-type condensing units, the hot gas line between the compressor and the condenser coil (in the cooling mode) is short enough so that any pressure drop in the line is ignored. If the condensing pressure is measured at the liquid line from the condenser coil, the condenser pressure drop of 1.5 psig is added to the liquid line pressure reading to determine the condensing pressure of the refrigerant off the compressor.

- *Condensing temperature.* Using a temperature-pressure chart, such as that shown in Figure 3-2, the condensing temperature of the refrigerant vapor is determined by finding the temperature equivalent of the pressure for the particular refrigerant in the system.

The total amount of liquid subcooling is equal to the condensing temperature minus the physical temperature of the liquid refrigerant. The physical temperature of the liquid refrigerant leaving the condenser coil should be measured within 6 in. of the condenser coil outlet manifold. This is done to reduce the line temperature loss that will affect the test results. A digital thermometer with 0.1°F increments is recommended.

Until the marketing of high-efficiency units, the industry used a small range of subcooling temperatures. Upright condensers operated with subcooling between 18° and 20°F, and horizontal condensers operated between 12° and 15°F. However, high-efficiency systems have higher pressure drops through the condenser coil circuits and stacked condensers, and there are no standards to determine subcooling. Some manufacturers publish the amount of subcooling that their systems require. If that information is available, it must be followed.

SYSTEM PERFORMANCE

When the system is low on refrigerant and the amount of refrigerant in the system is not known, pressure gauges will only show if there is sufficient refrigerant in the system to contain refrigerant in liquid form. If the system has been off long enough to have the physical temperature of the system components at ambient temperature, and if the system contains refrigerant in liquid form, the pressure in the system will be equal to the pressure equivalent of the ambient temperature.

If the system only contains vapor, the gauge pressure will be below the pressure equivalent of the ambient temperature. In this case, the best procedure is to recover the remaining refrigerant vapor, pressurize the system with nitrogen for a leak test, evacuate the system, and weigh in the refrigerant charge.

When the system contains liquid refrigerant, the system can be properly charged even if the quantity in the system is not known. This is done by charging refrigerant into the system until the maximum temperature drop is obtained through the evaporator in the a/c unit or heat pump in the cooling mode. This procedure was discussed in Chapter 2. (Heat pumps are designed to operate at peak performance in the cooling mode, and the refrigerant charge is based on the cooling mode requirement. In the heating mode, operating pressures usually are not excessive unless the unit is operating with an outdoor temperature higher than 65°F.) This difference in temperature may not be the desired amount for best results in the conditioned area. Adjustment for best results will be covered in another chapter.

If the system contains liquid refrigerant and the system passes a moisture-acid test, removing the refrigerant is not necessary. Instead, add refrigerant and record the difference in temperature of the air in the evaporator. Keep adding refrigerant until the peak difference in temperature is obtained. After the system has operated long enough for the temperature in the evaporator to stabilize, add refrigerant in 4 oz increments. Allow the system to stabilize after each increment addition.

As system efficiency improves, the difference in temperature through the evaporator will increase. By recording the beginning difference in temperature produced in the evaporator, the base temperature is established. Record the quantities added and the difference in temperature produced. Continue adding refrigerant each time the previous addition produces a rise in the difference in temperature across the evaporator.

In the beginning of the procedure, each addition of refrigerant may result in a 2°F or more change in the difference in temperature. When the charge quantity nears the peak performance, the difference in temperature change will be less per 4 oz increment. When this occurs, reduce the refrigerant charge increments to 2 oz per charge. When the peak performance is reached, any additional refrigerant will cause the difference in temperature to drop. When this situation is reached, remove the quantity of refrigerant that was added plus a quantity equal to the amount that would be in liquid form in the charging hoses used in the operation. For example, a 5-ft long hose filled with R-22 contains 0.4 oz per foot or 2 oz of refrigerant. This amount will be drawn into the system when the gauges are emptied for removal. **Do not overcharge the system.**

Systems using capillary tubes or restrictors must be critically charged, which means that the charge must be within 0.5 oz. Therefore, a short difference in temperature peak will be recorded. Systems using TXVs have a wider charge tolerance of +/-4 oz of refrigerant, and a broad peak will be recorded.

The system performance curve in Figure 3-4 is an example of the temperature difference across the evaporator as refrigerant is added into the system. Each addition of the refrigerant results in an increase in the temperature difference. This continues until the peak difference in temperature is reached and additional refrigerant causes a reduction.

After the system is properly charged, subcooling should be measured and recorded

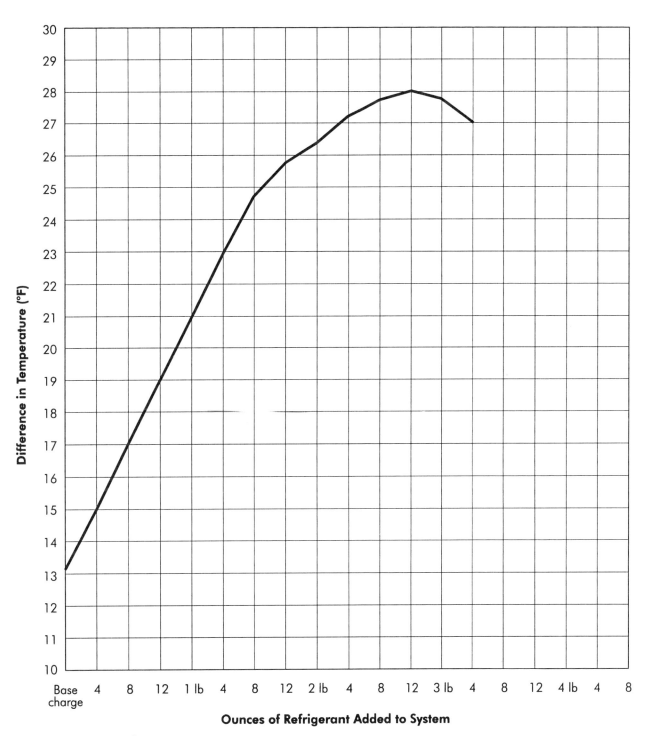

Figure 3-4. System performance curve

on the service record. This information can be used for future troubleshooting.

QUANTITY OF REFRIGERANT CHARGE

All manufacturers are required by ARI standards to list the type and quantity of refrigerant on the rating plate of the unit or outdoor portion of the system. On a package unit, the entire charge is listed.

In the split system, the amount of normal operating charge usually is given for the sections marketed by the particular manufacturer. To determine the total charge, the amount of refrigerant in the liquid and vapor lines is calculated. Depending upon the type of refrigerant, each line will contain a certain weight per foot. For example, Figure 3-5 shows that a 1/4-in. outside diameter (od) liquid line in a system using R-22 contains 0.39 oz per foot. If the line were 25 ft long, the refrigerant charge for the liquid line would be 9.75 oz (25 ft x 0.39 oz per foot) for the entire line. If the suction line were 1/2-in. od and 25 ft long, the refrigerant charge for the vapor line would be 0.75 oz (25 ft x 0.03 oz per foot) for the entire line. The total charge for the 25-ft line set would be 10.5 oz (9.75 oz + 0.75 oz). This line refrigerant quantity must be added to the refrigerant quantity specified by the manufacturer for the indoor and outdoor sections for proper system operation.

REFRIGERANT LINE SIZE IN. O.D.	CONTENTS PER FOOT IN OUNCES			
	LIQUID LINE		SUCTION LINE	
	R-22	R-12	R-22	R-12
1/4"	.39	.54	.01	.01
3/8"	.48	.64	.02	.01
1/2"	1.12	1.28	.03	.02
5/8"	1.76	2.08	.05	.03
3/4"	2.08	3.04	.08	.06
7/8"	3.84	4.16	.10	.08

Figure 3-5. Refrigerant contents in ounces per foot of line

When the manufacturer's charge information is not available and the system contains only refrigerant in vapor form, the refrigerant in the system must be recovered. The system should then be charged with dry nitrogen and checked for leaks. The leaks should be repaired and the system evacuated.

To arrive at the peak operating charge, a base charge is put into the system. A base charge that supplies enough liquid refrigerant without overcharging the system can be found by figuring 1.5 lb of refrigerant for each horsepower of the compressor. A 3-hp basic charge would be 1.5 lb x 3 hp, or 4.5 lb of refrigerant. After this quantity of refrigerant is put into the system, the charging procedure is the same as outlined previously.

PROBLEMS

3.1. Define superheat.

3.2. Is superheat latent or sensible heat?

3.3. The superheat in an evaporator is the difference between what temperatures?

3.4. What determines the amount of superheat in an evaporator using capillary tubes or a flow restrictor?

3.5. When adjusting the refrigerant charge in a system, what is the maximum increment quantity of refrigerant that should be added?

3.6. Define subcooling.

3.7. What is the purpose of having subcooling in the system?

3.8. What is the condenser split?

3.9. All air conditioning systems and heat pumps operate using the same condenser split. True or false?

3.10. The condenser split of a heat pump is the same in both the heating and cooling modes. True or false?

3.11. When the proper subcooling is not listed by the manufacturer, how can it be determined?

3.12. If there is no information available covering the operating pressures, subcooling, or superheat, on what basis can the system be properly charged?

3.13. If the correct refrigerant charge cannot be determined, what is a safe starting charge quantity?

CHAPTER FOUR

Determining System Capacity

After adjusting the airflow through the system to either the ARI standard of 400 cfm per 12,000 Btuh cooling capacity or the cfm specified by the manufacturer, the second step is to obtain the correct refrigerant charge by any of the following methods:

* Operating suction and discharge pressures

* Superheat according to the manufacturer's specifications

* Subcooling according to the manufacturer's specifications (if the proper subcooling is known)

* Peak performance

* Recovering refrigerant, evacuating system, and adding calculated amount of refrigerant

Whichever method is used to correct the refrigerant charge, the airflow through the evaporator must then be rechecked and adjusted accordingly. Correcting the charge usually increases the operating capacity of the system, resulting in an increase in the difference in temperature of the air flowing through the evaporator. After the adjustment is made, the system's cooling capacity can be checked against the manufacturer's rating.

NET COOLING CAPACITY

Figure 4-1 shows the cooling capacities of various system capacity ratings. These capacities are the *net cooling capacities* of the systems. Net cooling capacity is the unit rated capacity, which consists of the total Btuh removed from a conditioned area.

The net cooling capacity of a system can be determined either by the net cooling capacity method or the gross cooling capacity method. The net cooling capacity method involves accurately setting the cfm through the evaporator. The gross cooling capacity method consists of determining the gross cooling capacity and subtracting the heat generated by the condenser fan motor and the motor-compressor assembly. If the system does not have any form of auxiliary heat that can be used to determine the airflow through the evaporator and the air distribution is not easily accessible, the gross cooling capacity method is used.

NET COOLING CAPACITY METHOD

The net cooling capacity method involves measuring the airflow in cfm flowing through the evaporator and the entering and leaving wet bulb temperatures. The methods for determining the cfm flowing through the evaporator were covered in Chapter 2.

APPLICATION RATINGS			OUTDOOR TEMPERATURES °F*							
Condensing Unit Model Number	Evaporator Coil Model Number**	BTU/HR	80°	85°	90°	95°	100°	105°	110°	115°
18ECQ2	2ACQ1	Total Cooling	18950	18275	17600	16800	16175	15400	14550	13550
		Sensible Cooling	13350	13140	12900	12550	12325	12025	11750	11400
24ECQ2	2ACQ1	Total Cooling	26400	25525	24575	23400	22350	21150	19875	18500
		Sensible Cooling	18100	17670	17250	16850	16500	16150	15830	15510
30ECQ2	3ACQ3	Total Cooling	31750	31450	30800	30000	29000	27950	26975	26200
		Sensible Cooling	22825	22825	22800	22750	22500	21950	21250	20500
31ECQ	3ACQ3	Total Cooling	30000	29500	29000	28400	27600	26700	25500	23750
		Sensible Cooling	20300	20250	20200	20150	20100	20075	20050	20000
36ECQ4	3ACQ3	Total Cooling	37700	36400	35200	34200	32700	31500	30250	29000
		Sensible Cooling	24900	24200	23850	23500	23100	22800	22400	22000
37ECQ	3ACQ3	Total Cooling	36000	35800	35100	34600	32600	31200	29450	26600
		Sensible Cooling	24250	24550	24750	24750	24450	23700	22600	21250
42ECQ1	4ACQ2	Total Cooling	43800	43100	42300	41000	40050	38700	37100	34750
		Sensible Cooling	29550	25900	29450	29350	29200	28900	28250	26970
48ECQ2	4ACQ2	Total Cooling	46000	45900	45750	45000	44350	42550	40000	37250
		Sensible Cooling	30400	31100	31600	31850	31500	30850	29900	28700
60ECQ1	5ACQ1	Total Cooling	56500	55750	54900	54000	52700	51700	50500	48650
		Sensible Cooling	36350	36500	36650	36800	36350	35900	35450	35000

*All values based on 80° db/67° wb Return Air and Rated Evaporator Air Flow.
**For alternate evaporator coils, consult factory

Figure 4-1. System capacity ratings (Courtesy, Bard Manufacturing Co.)

To use this method, a psychrometric chart (see Figure 1-2) is necessary. Recall that on the psychrometric chart, the scale to the left of the wet bulb temperature curve indicates the enthalpy, or total heat content of a pound of air at its saturation temperature. For example, if the wet bulb temperature of the air entering the evaporator is 65°F, each pound of air holds 30.06 Btu. If the air passes through the evaporator and the wet bulb temperature is reduced to 55°F, each pound of air will hold 23.4 Btu. This means that 6.66 Btu were removed from each pound of that air.

The next step is to determine how many pounds of air are flowing through the evaporator. As the air increases in temperature and/or humidity, the air expands and each pound of air occupies a larger volume. The volume changes with a change in temperature and/or humidity, but the weight remains the same. The cfm of air flowing through the evaporator must be converted to pounds of air per minute (divide the cfm by the cubic feet per pound as found on the psychrometric chart). On the psychrometric chart, the lines that slant upward to the left and are marked 12.5, 13.0, 13.5, and 14.0 show the specific volume of air in cubic feet per pound for a given temperature and/or humidity level.

For this example, the return air is at 80°F db and 65°F wb, which means a cross point between the 13.5 and the 14.0 lines showing specific volume on the psychrometric chart. If the cross point were midway between the two cubic feet per pound lines, the specific volume would be 13.75 ft³/lb; however, it is closer to the 14.0 line than the 13.5 line. Measuring the distance between these two lines with the 3/16-in. scale on a drafting scale shows the distance between the 13.5 and 14.0 lines is 7 increments, and the cross point is 2.5 increments from the 14.0 line. Divide the 2.5 increments by the 7 increments to show that the cross point is 35.70% of the distance between the lines, and the difference between the lines is 0.5 ft³/lb. Therefore, if the cross point is 35.70% (.357) of 0.50, the cross point is 0.17 below the 14.0 line or 13.83 ft³/lb.

To calculate specific volume, it is also possible to use the following formula:

$$ANSU = \frac{(NSU)(0.5 \text{ ft}^3/\text{lb})}{TSU}$$

where: ANSU (actual number of scale units) = the difference between the ft³/lb at the cross point and the larger ft³/lb. Subtracting the difference from the larger ft³/lb line results in the value in ft³/lb at the cross point.

NSU (number of scale units) = number of scale units that the cross point of the dry bulb and wet bulb lines is below the larger ft³/lb line.

TSU (total scale units) = number of scale units between the ft³/lb lines that are involved.

0.5 = difference between the specific volume lines is 0.5 ft³/lb.

The pounds of air per minute (air lb/min) flowing through the evaporator are then determined by dividing the cfm flowing through the evaporator by the ft³/lb. For example, assume that 1200 cfm is moving through the evaporator. The air at 80°F db and 66.5°F wb or 50% relative humidity has a specific volume of 13.8 ft³/lb. By dividing the cfm by the specific volume, the air lb/min flowing through the evaporator can be determined. Therefore, 86.95 air lb/min multiplied by 6.65 Btu removed from each pound of air equals 578.21 Btu/min being picked by the evaporator. This is the enthalpy. Units are rated in Btuh, so 578.21 Btu/min x 60 minutes = 34,692.60 Btuh, the net cooling capacity of the system. Allowing for accuracy of the instruments used, the calculated rating should be within +/-10% of the manufacturer's rating.

Digital thermometers that read in 0.1°F increments will produce more accuracy in the net cooling capacity, because the psychrometric chart is not readable in 0.1°F increments. Figure 4-2 shows total heat content per pound of dry air with moisture to saturate it. The wet bulb temperatures and heat contents are listed in 0.1°F increments.

In Figure 4-2, the degree column on the left contains the wet bulb temperatures in whole numbers. The columns to the right of the degree column show the total heat of air in Btu/lb. The first column in this group (column heading of ".0") shows the heat content for the wet bulb temperatures. The columns to the right progress from 0.1° to 0.9°F. To use the previous example, the enthalpy for 65°F wb is 30.06 Btu/lb. If the thermometer reading were 65.1°F, the total heat would be 30.14 Btu/lb, etc. This table, or the use of digital thermometers, will produce more accurate results.

Gross Cooling Capacity Method

The gross cooling capacity of a cooling unit is the total amount of heat picked up in the evaporator and the compressor motor, which is then removed by the condenser coil. To determine the gross cooling capacity of the system, two factors must be determined: 1) the air in cfm flowing through the outdoor condenser coil; and 2) the temperature increase in that air.

Most manufacturers will list the air in cfm flowing through the condenser coil in their product literature. Figure 4-3 is an example of outdoor compressor unit specifications in which the air in cfm is listed under "Fan — Dia./CFM." For example, the 24HPQ2 unit has an 18-in. diameter fan and moves 1850 cfm through the outdoor coil. The 36HPQ4 unit has a 20-in. diameter fan and moves 2500 cfm through the outdoor coil.

To measure the temperature increase, temperatures of the entering and leaving air must be recorded. The entering air temperature is the ambient temperature surrounding the unit. Using a sling psychrometer, the air temperature is taken on all sides of the unit that have an air opening. The readings should be taken from a distance of 12 in. or more from the hot condenser surface to reduce the effect of radiating heat.

The temperature of the air leaving the unit needs to be an average of the readings taken over the area of the air outlet grille. On vertical discharge condensing units, such as the one shown in Figure 4-4, the thermometers are located on the top of the fan discharge grille. A reading should be taken above each condenser inlet face. For the

Wet Bulb Temperature F	TENTHS OF A DEGREE									
	.0	.1	.2	.3	.4	.5	.6	.7	.8	.9
35	13.01	13.05	13.10	13.14	13.18	13.23	13.27	13.31	13.35	13.40
36	13.44	13.48	13.53	13.57	13.61	13.66	13.70	13.75	13.79	13.83
37	13.87	13.91	13.96	14.00	14.05	14.09	14.14	14.18	14.23	14.27
38	14.32	14.37	14.41	14.46	14.50	14.55	14.59	14.64	14.68	14.73
39	14.77	14.82	14.86	14.91	14.95	15.00	15.05	15.09	15.14	15.18
40	15.23	15.28	15.32	15.37	15.42	15.46	15.51	15.56	15.60	15.65
41	15.70	15.75	15.80	15.84	15.89	15.94	15.99	16.03	16.08	16.13
42	16.17	16.22	16.27	16.32	16.36	16.41	16.46	16.51	16.56	16.61
43	16.66	16.71	16.76	16.81	16.86	16.91	16.96	17.00	17.05	17.10
44	17.15	17.20	17.25	17.30	17.35	17.40	17.45	17.50	17.55	17.60
45	17.65	17.70	17.75	17.80	17.85	17.91	17.96	18.01	18.06	18.11
46	18.16	18.21	18.26	18.32	18.37	18.42	18.47	18.52	18.58	18.63
47	18.68	18.73	18.79	18.84	18.89	18.95	19.00	19.05	19.10	19.16
48	19.21	19.26	19.32	19.37	19.43	19.48	19.53	19.59	19.64	19.70
49	19.75	19.81	19.86	19.92	19.97	20.03	20.08	20.14	20.19	20.25
50	20.30	20.36	20.41	20.47	20.52	20.58	20.64	20.69	20.75	20.80
51	20.86	20.92	20.97	21.03	21.09	21.15	21.20	21.26	21.32	21.38
52	21.44	21.50	21.56	21.62	21.67	21.73	21.79	21.85	21.91	21.97
53	22.02	22.08	22.14	22.20	22.26	22.32	22.38	22.44	22.50	22.56
54	22.62	22.68	22.74	22.80	22.86	22.92	22.98	23.04	23.10	23.16
55	23.22	23.28	23.34	23.41	23.47	23.53	23.59	23.65	23.72	23.78
56	23.84	23.90	23.97	24.03	24.10	24.16	24.22	24.29	24.35	24.42
57	24.48	24.54	24.61	24.67	24.74	24.80	24.86	24.93	24.99	25.06
58	25.12	25.19	25.25	25.32	25.38	25.45	25.52	25.58	25.65	25.71
59	25.78	25.85	25.92	25.98	26.05	26.12	26.19	26.26	26.32	26.39
60	26.46	26.53	26.60	26.67	26.74	26.81	26.87	26.94	27.01	27.08
61	27.15	27.22	27.29	27.36	27.43	27.50	27.57	27.64	27.71	27.78
62	27.85	27.92	27.99	28.07	28.14	28.21	28.28	28.35	28.43	28.50
63	28.57	28.64	28.72	28.79	28.87	28.94	29.01	29.09	29.16	29.24
64	29.31	29.39	29.46	29.54	29.61	29.69	29.76	29.84	29.91	29.99
65	30.06	30.14	30.21	30.29	30.37	30.45	30.52	30.60	30.68	30.75
66	30.83	30.91	30.99	31.07	31.15	31.23	31.30	31.38	31.46	31.54
67	31.62	31.70	31.78	31.86	31.94	32.02	32.10	32.18	32.26	32.34
68	32.42	32.50	32.59	32.67	32.75	32.84	32.92	33.00	33.08	33.17
69	33.25	33.33	33.42	33.50	33.59	33.67	33.75	33.84	33.92	34.01
70	34.09	34.18	34.26	34.35	34.43	34.52	34.61	34.69	34.78	34.86
71	34.95	35.04	35.13	35.21	35.30	35.39	35.48	35.57	35.65	35.74
72	35.83	35.92	36.01	36.10	36.19	36.29	36.38	36.47	36.56	36.65
73	36.74	36.83	36.92	37.02	37.11	37.20	37.29	37.38	37.48	37.57
74	37.66	37.76	37.85	37.95	38.04	38.14	38.23	38.33	38.42	38.52
75	38.61	38.71	38.80	38.90	38.99	39.09	39.19	39.28	39.38	39.47
76	39.57	39.67	39.77	39.87	39.97	40.07	40.17	40.27	40.37	40.47
77	40.57	40.67	40.77	40.87	40.97	41.08	41.18	41.28	41.38	41.48
78	41.58	41.68	41.79	41.89	42.00	42.10	42.20	42.31	42.41	42.52
79	42.62	42.73	42.83	42.94	43.05	43.16	43.26	43.37	43.48	43.58
80	43.69	43.80	43.91	44.02	44.13	44.24	44.34	44.45	44.56	44.67
81	44.78	44.89	45.00	45.12	45.23	45.34	45.45	45.56	45.68	45.79
82	45.90	46.01	46.13	46.24	46.36	46.47	46.58	46.70	46.81	46.93
83	47.04	47.16	47.28	47.39	47.51	47.63	47.75	47.87	47.98	48.10
84	48.22	48.34	48.46	48.58	48.70	48.83	48.95	49.07	49.19	49.31
85	49.43	49.55	49.68	49.80	49.92	50.05	50.17	50.29	50.41	50.54

Figure 4-2. Total heat content (enthalpy) of air in Btu/lb of dry air (Courtesy, ASHRAE Guide and Data Book, 1963)

SPECIFICATIONS • Split Heat Pump Systems

OUTDOOR COMPRESSOR UNIT

MODEL	18HPQ2	24HPQ2	30HPQ4	36HPQ4	36HPQ4-3†	42HPQ	42HPQ2-3†	48HPQ2	48HPQ2-3†	60HPQ4	60HPQ4-3†
Electrical Rating — 60 HZ	230/208 1-Ph	230/208 1-Ph	230/208 1-Ph	230/208 1-Ph	230/208 3-Ph	230/208 1-Ph	230/208 3-Ph	230/208 1-Ph	230/208 3-Ph	230/208 1-Ph	230/208 3-Ph
Operating Voltage Range	197-253V	197-253V	197-253V	197-253V	187-253V	197-253V	187-253V	197-253V	187-253V	197-253V	187-253V
*Minimum Circuit Ampacity	15A	18A	24A	34A	22A	30A	18A	33A	24A	40A	27A
**Delay Fuse Max	20A	30A	40A	50A	35A	50A	25A	50A	40A	60A	45A
Total Unit Amps	11.8	14.8	19.6	27.6	17.6	24.8	13.8	26.8	19.8	32.8	21.8
Compressor	PSC	PSC	PSC	PSC	3-Phase	PSC	3-Phase	PSC	3-Phase	PSC	3-Phase
Volts	230/208	230/208	230/208	230/208	230/200	230/208	230/208	230/208	230/200	230/208	230/200
Name Plate Amps	10	13	18	26	16	22	11	24	17	30	19
Lock Rotor Amps	43.3	64	75.8	108	82	110	74	115	93	180	158
Crankcase Heat	Capacitor Type	Capacitor Type	Immersion	Immersion	Immersion	Capacitor Type	Wraparound	Immersion	Immersion	Immersion	Immersion
Fan Motor & Condenser											
Fan Motor — HP/RPM	1/5-950	1/5-950	1/5-1075	1/5-1075	1/5-1075	1/3-825	1/3-825	1/3-825	1/3-835	1/3-825	1/3-825
Fan Motor — AMPS	1.5A	1.5A	1.6A	1.6A	1.6A	2.8A	2.8A	2.8A	2.8A	2.8A	2.8A
Fan — Dia./CFM	18"/1850	18"/1850	20"/2450	20"/2500	20"/2500	24"/3600	24"/3600	24"/3600	24"/3400	24"/3400	24"/3400
Face Area Sq. Ft./Row/Fins/Inch	3.75/3/14	3.75/3/14	5.04/3/14	5.04/3/14	5.04/3/14	7.7/2/14	7.7/2/14	7.7/2/14	7.7/2/14	7.7/3/12	7.7/3/12
Refrigerant Control	Capillary	Capillary	Capillary	Capillary	Capillary	Capillary	Capillary	Capillary	Capillary	Capillary	Capillary
Shipping Weight Lbs.	170	180	210	220	220	261	258	275	272	291	283

*For sizing of CU Wire   **Or Type HACR Circuit Breaker   †Also available in 460V (460V not UL listed)

Figure 4-3. Outdoor unit specifications (Courtesy, Bard Manufacturing Co.)

condensing unit shown in Figure 4-4, three readings would be taken and an average temperature calculated. If the unit were a wrap-around type, four readings, equally spaced around the top grille, would be taken and averaged.

In both units, the condenser fan is a draw-through fan with a fair amount of air mix leaving the fan discharge grille. If the unit were a blow-through discharge design, the average air temperature would be more difficult to obtain because of the wide variety of air quantities coming from each section of the propeller fan blade.

Figure 4-5 shows a horizontal discharge condensing blow-through unit. In this case, more readings must be taken to arrive at the average air temperature. The face of the discharge grille is divided into equal size segments, and the temperature from each segment is recorded and averaged. The airflow in cfm varies widely over the face of the coil. The center of the fan hub and the four corners of the coil have reduced quantities of air.

Figure 4-4. Vertical discharge condensing unit (Courtesy, Goettl Air Conditioning, Inc.)

After determining the average air temperature, the temperature rise of the air through the condenser coil is calculated. The average leaving air temperature (LAT) minus the

Figure 4-5. Horizontal discharge condensing unit (Courtesy, Bard Manufacturing Co.)

entering air temperature (EAT) equals the temperature increase (TD) of the air. The heat rejected from the condensing unit is determined by using the standard air heat content formula:

Btuh = (cfm)(TD)(1.08)

where: Btuh = sensible heat rejected from the condenser coil — no moisture is involved. All the heat energy rejected into the air will be reflected in the temperature increase.

cfm = amount of air traveling through the condenser coil (taken from manufacturer's specifications).

TD = temperature difference in the air.

1.08 = a constant used to convert cfm to cubic feet per pound and from pounds per minute to pounds per hour. It also contains the quantity of heat energy needed to change the temperature of each pound of air.

Multiply the cfm flowing through the condensing unit, which can be obtained from the manufacturer's specifications, by the temperature rise of the air flowing through the unit. Then multiply that product by the constant of 1.08 to obtain the gross cooling capacity of the system.

Motor Heat

The amount of heat added to the vapor as it passes through the compressor operation is practically 100% of the electrical energy needed to operate the motor compressor assembly and condenser fan motor. The small loss from the compressor shell by conduction is ignored. The formula used to calculate motor heat is as follows:

Motor heat, Btuh = (volts)(amps)(PF)(3.413 Btu/watt)

where: volts = voltage at the load side of the compressor contactor with the unit operating. The actual operating voltage rating must be used regardless of the manufacturer's voltage rating. If it is a three-phase unit, the voltage across each phase should be the same as the others. If a voltage difference of more than 3% is found, the voltage unbalance will cause amperage unbalance, increase motor winding temperatures, and shorten motor life. The utility supplying power should be contacted about this problem.

amps = total amperage draw of the condensing unit. This includes the amperage drawn by both the motor-compressor assembly and the condenser fan motor.

PF = power factor[1] of the entire outdoor condensing unit. The power factor of condensing units will vary between 86% and 94% depending upon the applied voltage and the load on the unit. An average of 90% or 0.9 is used in the formula. The average power factor is 0.9. For more accurate calculations, a "true watt" meter should be used.

Btu/watt = heat energy of electrical energy, which is 3.413 Btu per watt of electrical energy.

The system net cooling capacity, which is the total capacity of the evaporator, is then obtained by subtracting the motor heat input from the gross cooling capacity. Compare these results with the capacities listed in the manufacturer's literature. The inaccuracy of field instrumentation must allow a tolerance of +/-10%.

COOLING CAPACITY, STEP-BY-STEP

As already discussed, determining the net cooling capacity of an a/c system or heat pump in the cooling mode involves many steps. Figure 4-6 illustrates these different steps in a flowchart format.

To measure the net cooling capacity of an air conditioning unit, the basic test cfm must be established before further testing can be accomplished. The following paragraphs describe the steps in Figure 4-6.

- Step 1. Find the unit's airflow rated by ARI in cfm, or use the rate listed in the manufacturer's specification sheets. The manufacturer's cfm quantities for each unit will produce more accurate results. However, when the manufacturer's specifications are not available, use the ARI standard of 400 cfm per 12,000 Btuh of the unit cooling capacity at 95°F outdoor ambient temperature. This rate in cfm is to be used when testing the unit and has been classified as the desired rate. Record the cfm for the test.

- Step 2. Operate the auxiliary heating unit to determine the output of the unit in Btuh. The procedure to determine the output of an electrical or fossil fuel heating unit was discussed in Chapter 2.

- Step 3. Calculate the difference in temperature (TD) that the cfm and output (Btuh) should produce by using the ARI or manufacturer's rating for the cfm and the unit's output obtained in Step 2. Record this difference in temperature.

- Step 4. With the auxiliary heating unit operating, adjust the blower speed and/or the swing damper to obtain an actual operating difference in temperature that is the same as the desired difference in temperature. This establishes the test cfm to be used in the cooling mode test. This may not be the final difference in temperature to produce the best results for the local weather conditions. When testing operation in the cooling mode, cfm must be that of ARI or manufacturer.

- Step 5. With the unit operating in the cooling mode and the cfm at either ARI or manufacturer's rating, the total net cooling capacity of the unit (both sensible and latent) is determined. The total heat content of the air (Ht) is based on wet bulb temperature. The entering and leaving air wet bulb temperatures (EAT wb °F and LAT wb °F) are measured at the return air and supply air duct connections at the unit. These temperatures should be taken with digital thermometers that can read in 0.1°F increments. Once the wet bulb temperatures are recorded, find the total heat content of the air (Ht) using Figure 4-2, or use the psychrometric chart. The subtraction in this step will show how many Btu are removed for each pound of air flowing through the evaporator.

- Step 6. To determine the specific volume (Vs) of the entering air to the evaporator, the entering air dry bulb and wet bulb temperatures are recorded. Using the psychrometric chart, the relative humidity (RH) of the entering air is determined. When the RH point is located, it is used to determine the specific volume of the air at these conditions. The specific volume (ft^3/lb) is then recorded.

Unit Model No._____ Rated Btuh _____ Outdoor °F _____

1. Find ARI-rated cfm:

    $$\frac{\text{(Rated Btuh)(400 cfm)}}{\text{12,000 Btuh}}$$

    or manufacturer's rated cfm: _____cfm

2. A/c unit off, auxiliary unit on. Find Btuh output for electric or fossil fuel-powered heating unit.

    Electric = (volts)(amps)(3.413)(Efficiency)

    Fossil = (input)(efficiency)

    Output is _____Btuh

3. Find desired TD _____°F

    $$TD = \frac{\text{(Btuh output)(Efficiency)}}{\text{(cfm)(1.08)}}$$

4. With auxiliary unit operating, adjust cfm so that actual and desired TD are the same.

5. Auxiliary unit off, a/c unit on. Find difference in Ht of air flowing through evaporator.

    EAT wb °F _____ Ht _____Btu/lb

    LAT wb °F _____ Ht _____Btu/lb

    TD Ht _____Btu/lb

6. Find EAT Vs _____ft³/lb

7. Find airflow in lb/min flowing through evaporator.

    $$\frac{\text{cfm}}{\text{ft}^3/\text{lb}} = \text{lb/min}$$

8. Find total net cooling capacity.

    Ht, Btuh = (_____lb/min)(_____Ht, Btu/lb)(60)

Figure 4-6. Flowchart for calculating net cooling capacity, continued on next page

9. Compare Ht to manufacturer's rate. Ht should be within +/-5% of manufacturer's rating.

   Ht _____ Btuh

   Manufacturer's rate _____ Btuh

   $$\frac{\text{Manufacturer's rate}}{\text{Ht}} = \% \text{ of rate}$$

10. Find desired TD for local conditions.

    EAT db °F _____, EAT wb °F _____

    RH% ____. Desired TD_____

11. Adjust cfm so that actual TD is same as desired TD.

    If neither an electric nor fossil fuel-powered auxiliary heating unit is available, actual cfm can be determined by any of the following methods:

12. Find desired fpm flowing through the duct (to supply the rated cfm for Step 1).

    $$\text{Avg. fpm} = \frac{\text{Rated cfm}}{\text{Duct area, ft}^2}$$

13. From velocity pressures table:

    Desired fpm _____ = avg. Vp _____ in. wc

14. Using static pressure gauge and pitot tube, adjust cfm to match actual avg. Vp with desired Vp.

15. With rated cfm, return to Step 5.

16. Using static pressure gauge, measure supply and return static pressure; set ESP to obtain desired cfm. Use manufacturer's literature.

17. With rated cfm, return to Step 5.

18. Use flowhood to measure total cfm from all supply outlets.

19. Adjust fan speed or swing damper so that total cfm matches desired cfm.

20. With rated cfm, return to Step 5.

Figure 4-6. Continued from previous page

- Step 7. Divide the cfm from Step 1 by the specific volume (Vs) in ft³/lb from Step 6 to obtain the air lb/min flowing through the evaporator.

- Step 8. Multiply the air lb/min from Step 7 by the Btuh removed per pound (Ht, Btu/lb) in Step 5, to obtain the amount of heat energy in Btu/min. The Btu/min multiplied by 60 minutes per hour will give the actual total net cooling capacity (Ht, Btuh) of the unit.

- Step 9. Compare the actual net cooling capacity to the manufacturer's rating at the outdoor temperature. The unit capacity in Btuh should be within +/-5% of the manufacturer's rating in cfm. If the 400 cfm per 12,000 Btuh rating is used, the unit capacity in Btuh may be as much as 10% below the manufacturer's rating. This is because some manufacturers rate their units using as much as 440 cfm per 12,000 Btuh, which is 10% over ARI rating. If the actual rating is more than 5% above the manufacturer's rating, the test should be repeated for more accuracy. If the actual rating is more than 20% below the manufacturer's rating, the refrigerant charge or refrigerant flow rate should be checked.

- Steps 10 and 11 will be covered in Chapter 5.

- Step 12. Divide the ARI rating or manufacturer's rating in cfm by the duct area (ft²) to find the desired average duct velocity (fpm).

- Step 13. Using the velocity pressures table in Figure 2-11, the desired average velocity pressure (Vp) in inches of water column (in. wc) is determined.

- Step 14. By measuring the duct velocity using a static pressure gauge and pitot tube, traversing the duct, and averaging the readings, the actual average velocity (Vp) is determined. The fan speed and/or the swing damper are adjusted to match the actual average duct velocity with the desired average duct velocity. This means the desired cfm is flowing through the duct.

- Step 15. When the actual and desired fpm are the same, the unit is operating at rated cfm and Steps 5 through 9 are complete. Compare the unit total cooling capacity to the rated total cooling capacity. This test requires more time than testing with an auxiliary heating unit, but it is required when no auxiliary heating unit is available.

- Step 16. A static pressure gauge with two taps, one in the supply and one in the return, is needed to set the manufacturer's rated cfm (see Figure 2-16). The manufacturer's specifications must supply the cfm flowing through the unit with a range of external static pressures (ESP) — supply and return duct pressures.

- Step 17. When the rated cfm is attained, Steps 5 through 9 are completed and the unit total net cooling capacity may be found.

- Step 18. The use of a flowhood is the easiest method to use when no auxiliary heating unit is available. The cfm from each supply outlet is measured, and the total of all outlets is calculated.

- Step 19. The total cfm is then adjusted to match the desired cfm by fan speed and/or swing damper adjustment.

- Step 20. With the cfm adjusted to the rated cfm, Steps 5 through 9 are completed to find the unit net cooling capacity.

HEATING CAPACITY

The cfm flowing through the heat pump was established in the cooling mode. No further airflow adjustment is made in the heating mode unless the cooling cfm results in a leaving heating air temperature over 105°F when the outdoor temperature is 65°F. Then, a two-speed blower motor is required to adjust the cfm to the proper amount in each mode. This situation is unusual but could exist.

Figure 4-7 is a flowchart for calculating the rated gross capacity of a heat pump in the heating mode.

The following paragraphs describe the steps in Figure 4-7.

- Step 1. Find the ARI-rated cooling cfm of the heat pump or use the manufacturer's listed rate in cfm. When the manufacturer's specifications are not available, use the ARI standard of 400 cfm per 12,000 Btuh cooling capacity. This cfm is to be used when testing the unit in the heating mode and is the desired cfm.

- Step 2. Operate the auxiliary heating unit to determine the output in Btuh. The procedure to determine the output of an electrical or fossil fuel heating unit was discussed in Chapter 2.

- Step 3. Calculate the difference in temperature (TD) through the auxiliary heating unit. Calculating the TD using the ARI or manufacturer's cfm rating and the unit output should result in the actual TD that the unit should produce. Record the TD.

- Step 4. With the auxiliary heating unit operating, adjust the blower speed and/or the swing damper to obtain an actual operating difference in temperature that is the same as the desired difference in temperature. This establishes the cfm for the gross capacity heating (gch) test.

- Step 5. With the pump operating in the heating mode, the auxiliary heating unit is off and the operating difference in temperature may be established. The EAT and LAT dry bulb temperatures and the operating difference in temperature are then recorded. The total heat of the air through the heat pump is based on dry bulb temperatures. No humidity is involved — the heat energy transferred is only sensible heat. Therefore, the supply (SAT) and return (RAT) air temperatures are measured and the difference in temperature is calculated. These temperatures should be taken with digital thermometers that read in 0.1°F increments.

- Step 6. Using the standard air formula, calculate the gross capacity heating of the heat pump.

- Step 7. Compare the calculated gross capacity heating to the manufacturer's rated capacity at the outdoor temperature measured during the test.

If the manufacturer's specifications are known, a comparison can be made. The determined capacity should be within +/-5% of the manufacturer's rating for the outdoor ambient temperatures encountered if the manufacturer's cfm ratings are used. If the ARI standard of 400 cfm per 12,000 Btuh unit cooling capacity is used, the test rating may be as much as 10% below the manufacturer's rating. If the calculated capacity is more than 5% above the manufacturer's rating, the tests should be performed again using more accuracy in reading the instruments. If the calculated capacity is less than 90% of the manufacturer's rating, the a/c system should be checked. This will be discussed in Chapter 6.

If no auxiliary heat is installed with the heat pump and the instrument method is used to find the rated cfm, refer to the procedures outlined in Steps 8 through 16 in Figure 4-6.

> Unit Model No._____ Rated Btuh _____ Outdoor °F _____
>
> 1. Find ARI-rated cfm:
>
> $$\frac{(\text{Rated Btuh})(400 \text{ cfm})}{12{,}000 \text{ Btuh}}$$
>
>   or manufacturer's rated cfm: _____cfm
>
> 2. A/c unit off, auxiliary unit on. Find Btuh output for electric or fossil fuel-powered heating unit.
>   Electric = (volts)(amps)(3.413)
>   Fossil = (input)(efficiency)
>   Output is _____Btuh
>
> 3. Find desired TD _____°F
>
> $$TD = \frac{(\text{Btuh output})(\text{Efficiency})}{(\text{cfm})(1.08)}$$
>
> 4. With auxiliary unit operating, adjust cfm so that actual and desired TD are the same.
>
> 5. Auxiliary unit off, a/c unit on. Find the difference in Ht of air flowing through evaporator.
>   SAT db °F _____
>   RAT db °F _____
>   TD _____ °F
>
> 6. Find gross capacity heating:
>   Btuh = (Rated cfm)(TD)(1.08)
>   Btuh = _____
>
> 7. Compare gross capacity heating to manufacturer's rated capacity at outdoor temperature measured during test. Capacity should be within +/-5% of manufacturer's rating.
>   Manufacturer's rate _____Btuh
>   gross capacity heating _____Btuh_____%

Table 4-7. Flowchart for calculating gross capacity of a heat pump in heating mode

PROBLEMS

4.1. What are the five different methods that can be used to obtain the correct refrigerant charge in an air conditioning system or heat pump?

4.2. What are the two different methods used to obtain the net cooling capacity of an air conditioning system or heat pump?

4.3. What is the formula for converting cubic feet per minute to pounds per hour?

4.4. When determining the gross cooling capacity of an air conditioning system or heat pump, both the dry bulb and wet bulb temperatures are needed. True or false?

4.5. On a vertical discharge complete wrap-around condensing unit, what is the minimum number of discharge temperature readings required?

4.6. On a horizontal discharge blow-through type condensing unit, the most accurate leaving air temperature reading is in the middle of the condenser face area. True or false?

4.7. What is the formula for finding the heat in Btuh removed from or put into air?

4.8. What is the formula for finding motor heat input?

4.9. Define power factor.

NOTES

[1] Power factor may be defined as the true power in watts divided by the apparent power in volt-amperes. It is also the cosine of the phase angle between the voltage applied to a load and the current passing through it.

CHAPTER FIVE
Adjusting for Proper Local Performance

When adjusting fossil fuel-powered heating units for peak performance, the first step is to adjust the unit for the rated input in Btuh. Gas- and oil-fired heating units transfer heat energy from materials that contain potential heat energy to air flowing through the unit. The heat energy is released by the process of combustion. The released energy is absorbed by the heat exchanger, which transfers the heat energy to the moving air over the surface of the heat exchanger. Top performance is achieved by ensuring the unit is receiving the correct input in Btuh.

Electric heating units have established inputs according to the electrical energy that the elements are designed to handle. Therefore, the electrical energy is preset and may not be adjusted in the field.

A/c systems and heat pumps are also heat energy-transfer systems. Their input depends on the quantity of sensible and latent heat in the air and the air quantity flowing through the evaporator.

This chapter will discuss the elements necessary for Steps 10 and 11 in Figure 4-6.

COMFORT ZONE
Many tests have been performed to determine a comfort zone, which is the range of temperature and humidity at which the greatest number of people are comfortable. This range of temperatures and humidity is plotted on the psychrometric chart in Figure 5-1. The area outlined in heavy lines is the range of dry bulb temperatures and relative humidity at which all test groups expressed some degree of comfort. The greatest number of people were most comfortable at 80°F dry bulb and 66.5°F wet bulb or 50% relative humidity. As a result, these conditions in the conditioned area were adopted by the air conditioning industry as part of the design standards. The objective of fine tuning the cooling system is to maintain conditions that will satisfy the greatest number of people.

With the correct amount of air flowing through the evaporator of an a/c system or heat pump in the cooling mode, the unit will produce the desired temperature drop in the air. This will remove moisture (latent heat) and lower the temperature (sensible heat) to the desired inside conditions within the unit's range of operating conditions.

TEMPERATURE DROP
The first step in obtaining the desired drop in temperature through the evaporator is to find the amount of heat energy in the air of the occupied area. The amount of heat energy, both sensible (dry bulb) and latent (wet bulb),

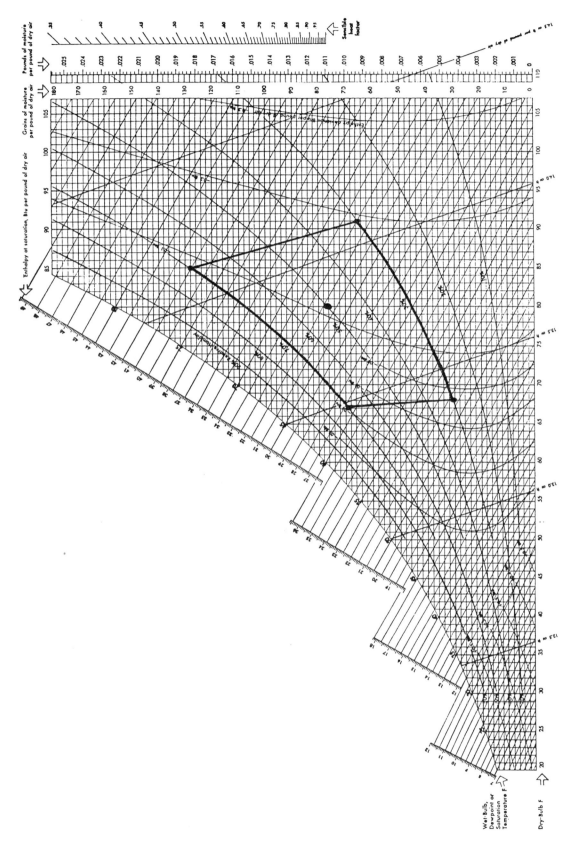

Figure 5-1. Psychrometric chart showing comfort zone (Courtesy, Carrier Corporation)

is measured by using a sling psychrometer in the air entering the evaporator. The relative humidity can be obtained with the dry bulb and wet bulb readings and the psychrometric chart. For example, consider that the design conditions for air conditioning systems are 80°F dry bulb and 66.5°F wet bulb. On the psychrometric chart, the vertical 80°F line and the 66.5°F wet bulb line cross at the 50% relative humidity curve. The conditions in the area of 80°F dry bulb, 66.5°F wet bulb, and 50% relative humidity are then plotted on the chart showing air flowing through the evaporator, Figure 5-2.

The horizontal lines in Figure 5-2 show dry bulb temperatures, indicated by the right-hand scale, and the slanted lines show relative humidity. The bottom scale is the drop in temperature through the evaporator that will result from the various conditions of the air entering the evaporator. If the 80°F dry bulb and 50% relative humidity measurements are plotted on the chart, the point at which the lines cross (Point A) is at the 20°F temperature drop (vertical line). The problem is that the 20°F drop in temperature is usable in a very limited range of air conditions. The temperature that a cooling unit produces varies with the heat content of the air.

A refrigeration system will only transfer a quantity of heat energy set by the system design. It will handle the heat load by adjusting the drop in temperature produced according to the ratio of sensible heat to latent heat. A change in the sensible heat quantity and/or the latent heat quantity will result in a change in the drop in temperature in the air flowing through the evaporator.

In Figure 5-3, the temperature change in the air that is passing between the fins of the evaporator is plotted. As the air passes between the fins, the air temperature is lowered below its dew point and moisture condenses out of the air. The closer to the fin that the air travels, the lower the resulting air temperatures. The air passing through the center of the space between the fins is not cooled as much, and it dilutes the air leaving the fin surfaces. This raises the temperature of the leaving air.

Fin coils are not 100% efficient, because the amount of air that passes through the evaporator is only partially treated. This portion of untreated air is called the *bypass factor*, and it varies from 5% for 13 fin-per-inch coils to 3% for higher-efficiency evaporators. When plotting the performance line on the psychrometric chart, the horizontal dry bulb temperature line indicates a sensible temperature change to 95% relative humidity; the line then follows the sensible and latent heat removal curve of 95%.

The temperature reduction line is plotted from the entering air temperature of 80°F dry bulb (Point A, Figure 5-3) to the 95% relative humidity curve (Point B, Figure 5-3) to the leaving air temperature of 60°F (Point C, Figure 5-3). If sensible heat is added to the air and the dry bulb temperature is raised to 85°F without adding moisture (Point D, Figure 5-3), the relative humidity is decreased to 43%. The required drop in temperature of the air through the evaporator is raised to 23°F (Point B, Figure 5-3). The moisture load (Hl) on the unit remains the same, and the relative humidity decreases. The system automatically adjusts the heat removal rate between sensible heat load and latent heat load to handle the increase in sensible heat load. The system increases the drop in temperature through the evaporator from 20° to 23°F.

If the conditions of the air in the conditioned area are 85°F dry bulb and 43% relative humidity, the airflow through the evaporator would have to be adjusted to produce a drop in temperature of 23°F. This temperature change would result in the desired design standard of 80°F dry bulb and 50% relative humidity. Removal of moisture from the air requires more unit capacity than removal of sensible heat. The drop in temperature through the evaporator will be affected more by a change in relative humidity than by a change in sensible (dry bulb) temperature.

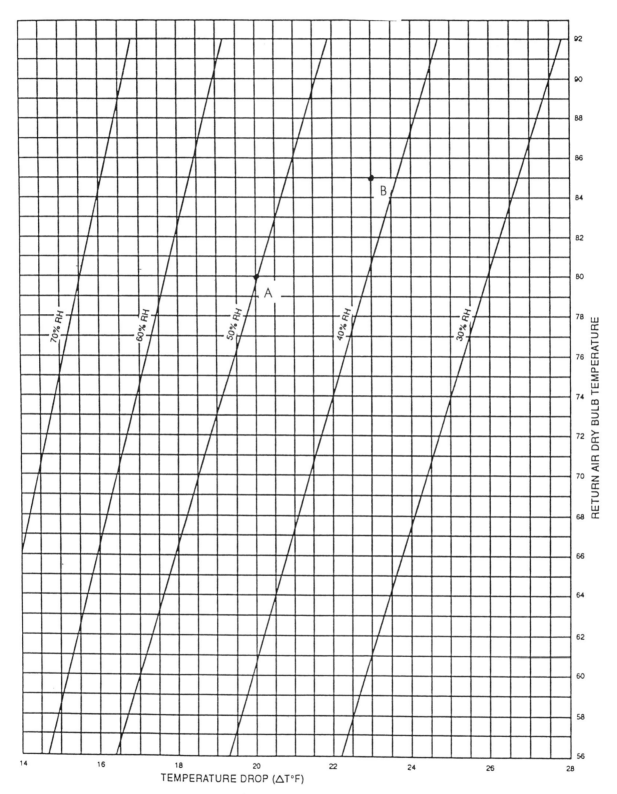

Figure 5-2. Temperature drop of air flowing through evaporator

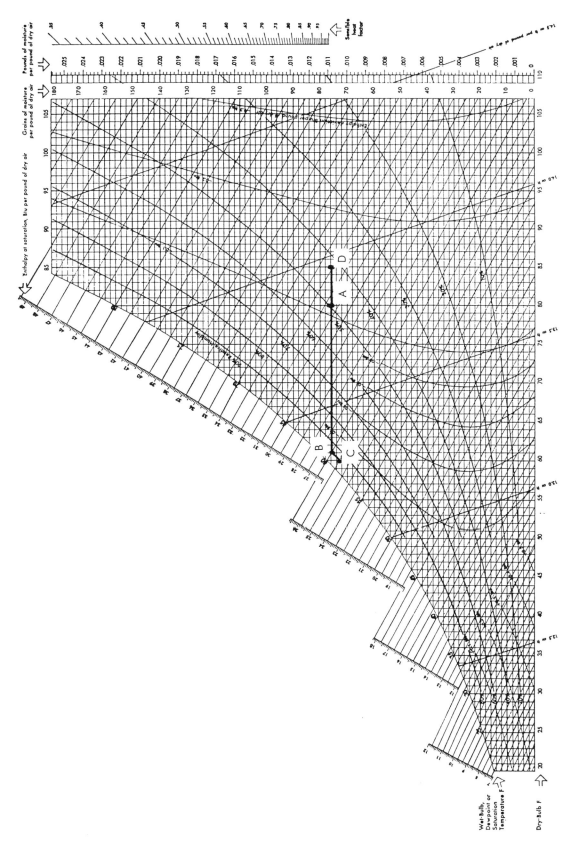

Figure 5-3. Psychrometric chart showing the temperature change in air passing through fins of the evaporator (Courtesy, Carrier Corporation)

Adjusting for Proper Local Performance 61

If the relative humidity of the air is increased from 50% to 60% but the sensible temperature (dry bulb) is kept at 80°F, the portion of the temperature difference removed is reduced and the portion of latent heat removed is increased. The total temperature reduction is almost 18°F, because the increase in the amount of latent heat removed has reduced the capability of the unit to remove sensible heat.

To determine the correct drop in temperature in the air flowing across the evaporator, perform the following steps:

1. Measure the dry bulb and wet bulb temperatures of the air entering the evaporator.

2. At the same time, measure the dry bulb temperature of the air leaving the evaporator.

3. Subtract the leaving dry bulb temperature from the entering dry bulb temperature to obtain the actual temperature difference across the evaporator.

4. Using the psychrometric chart in Figure 5-1, determine the relative humidity of the entering air.

5. Using the dry bulb temperature and relative humidity of the entering air, Figure 5-2, determine the desired drop in temperature through the evaporator.

6. Adjust the airflow in cfm accordingly to obtain the desired drop in temperature through the evaporator. This procedure will be explained in the paragraphs that follow.

If the measured drop in temperature through the evaporator is less than desired, reduce the amount of air flowing through the evaporator until the two drops in temperature match. If the measured drop in temperature is more than desired, increase the air flowing through the evaporator.

ADJUSTING TEMPERATURE DROP

When blower assemblies in heating units were belt-driven, precise settings of the drop in temperature through the evaporator were possible by adjusting the motor pulley to within 1/2 revolution. With direct-drive blower motors, the motor speed is adjusted. If a single speed blower motor is supplied with the unit, a multispeed motor may have to be substituted.

By operating the unit in the cooling mode, the resulting drop in temperature for each speed is obtained. If the desired drop in temperature is obtained when operating on one of the motor speeds, no further adjustment is necessary. If the desired temperature adjustment falls between the motor speeds, a means of adjustment is easily added to the blower scroll.

Figure 5-4 shows a direct-drive blower housing with an adjustment plate located on the "dead" side of the blower assembly (the dead side is the side opposite the motor). By installing an adjustment plate on the inlet side of the blower housing, airflow can be adjusted between the various speeds of the blower motor.

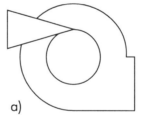 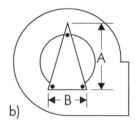

Figure 5-4. Airflow adjustment plate

To make the adjustment, the adjustment plate is moved as far off the evaporator inlet venturi as possible to reduce the flow restriction of the air into the blower scroll, Figure 5-4a. The system is operated in order to find the drops in temperature obtained with the motor speeds. When the motor speeds are found and the desired drop in temperature is obtained between these speeds, the unit is operated on the blower that produces the lower drop in temperature.

The adjustment plate is lowered across the blower inlet venturi until the desired drop in temperature is obtained, Figure 5-4b. The adjustment plate is then secured with three sheet metal screws to the side of the scroll housing. There are three screw locations, one at each point of the triangle. The top screw is the pivot point, and the other two screws anchor the bottom of the plate. Three screws are necessary to prevent the plate from rattling. Use No. 6, half-inch, sheet metal screws to keep the adjustment plate in place. Larger screws will interfere with the blower wheel.

Blower scrolls vary in size, so the venturi opening into the blower wheel varies in size. The easiest way to remember the size of the adjustment plate is as follows:

- The height from the middle of the bottom to the top of the plate is 2 in. longer than the outside diameter of the venturi opening to the wheel. This is measured at the front of the venturi opening, in line with the flat side of the blower scroll, Dimension A, Figure 5-4b.

- The width of the base of the plate is the outside diameter of the front of the venturi opening, in line with the side of the blower scroll, Dimension B, Figure 5-4b.

PROBLEMS

5.1. Define comfort zone.

5.2. At what conditions of air are the most people comfortable?

5.3. Which portion of the air conditioning load has the greater effect on the drop in temperature through the evaporator — latent heat or sensible heat?

5.4. Using the psychrometric chart and the chart showing the drop in temperature of air through the evaporator, what is the desired drop in temperature for an EAT 78°F dry bulb and 67.5°F wet bulb?

5.5. As the air entering the evaporator is reduced in relative humidity, the load is reduced. Therefore, the drop in temperature of the air drops. True or false?

5.6. When the temperature of the air flowing through the evaporator is lowered below the dew point, the air leaves the evaporator in a saturated condition. True or false?

5.7. If the drop in temperature of the air through the evaporator is too low, it is because the system is not removing enough heat from the air. To correct this, the airflow through the evaporator should be increased. True or false?

5.8. The size of a blower adjustment plate is based on what dimensions for the height and width of the plate?

CHAPTER SIX
Troubleshooting

As stated in earlier chapters, problems in an air conditioning system may be related to either refrigerant quantity or refrigerant flow rate. If the refrigerant quantity and flow rate are correct, the system should perform as it is designed. However, other factors can affect system operation even though the refrigerant charge is correct.

Troubleshooting discussions in this book have all been based on systems in which the difference in elevation between the high and low sides of the system is less than 20 ft and the lines between the sections are subjected to the ambient temperatures surrounding the outdoor unit. In systems that are installed in locations where the difference in elevation between the indoor and outdoor sections is more than 20 ft and where the refrigerant lines go through areas in which the temperature is higher than the ambient temperature of the outdoor section, these factors will cause the system to operate at reduced capacity and affect all other pressure and temperature readings. This chapter covers ordinary problems, as well as unusual problems that affect system operation.

REFRIGERANT CHARGE
If the amount of subcooling is lower than that obtained from a previous reading, the system has lost refrigerant. Recover the remaining refrigerant from the system, pressurize the system with nitrogen, find the leak, repair it, and evacuate and recharge the system.

If the amount of subcooling is higher than that obtained from a previous reading, the system may have an overcharge of refrigerant, a flow restriction in the liquid line, or false pressures exist in the system.

OVERCHARGE OF REFRIGERANT
In systems using capillary tubes or restrictors, refrigerant may be flooding back to the compressor. In this case, head pressure and amperage draw will be higher than normal and superheat will be 0°F. In systems using thermostatic expansion valves, superheat will be close to normal, but high head pressure and amperage draw will be found.

FLOW RESTRICTION
The subcooling is high when the refrigerant is backing up in the condenser due to excessive resistance (pressure drop) in the liquid line. Less heat is picked up in the evaporator because of the reduced refrigerant flow. Because there is less load on the compressor, its discharge and amperage draw are reduced. A restriction can occur in the liquid line, liquid line filter drier, capillary tube(s), and/or capillary tube inlet screen, or it can

be due to a difference in elevation. In addition, the system can have the correct amount of refrigerant charge and have reduced flow rate even though there are no mechanical restrictions in the liquid line.

HIGHER-THAN-NORMAL AMBIENT TEMPERATURES

When the liquid line is subjected to higher-than-normal ambient temperatures, expansion of the liquid refrigerant before it enters the metering device will cause loss in capacity.

Air Handler in Attic

One example of a situation in which higher-than-normal ambient temperatures may occur is when the air handler is located in the attic, Figure 6-1. Consider this air handler to be a high-efficiency air conditioning system operating with 90°F air entering the outdoor condenser coil. The liquid line rises 10 ft from the outdoor section and travels across the attic for 25 ft to the air handler unit. Temperatures in the attic may reach 130°F. The condensing unit operates with a 20°F split, and there is 15°F subcooling at the condenser. This means that the vapor is condensing at 90°F plus the 20°F split, or 110°F. The liquid in the condenser coil is subcooled 15°F and leaves the condenser coil at 95°F. The liquid could lose 1°F through the upright portion of the line exposed to the outdoor temperature. If this 94°F liquid is exposed to 130°F attic temperature, the liquid could reasonably pick up 20°F. This destroys part of the subcooling effect and increases the amount of flash gas formed in the evaporator. This creates the same effect as a shortage of refrigerant, which reduces system capacity and increases the operating cost.

For the reasons just described, **liquid lines must be insulated if they pass through spaces or over surfaces that contain temperatures higher than the outdoor temperatures**. Attics are an example of higher-than-normal temperature areas. Other examples include the ceilings above boilers in commercial installations such as apartment buildings.

Roof Installation

A tar-coated flat roof of a commercial building is another example of an area in which

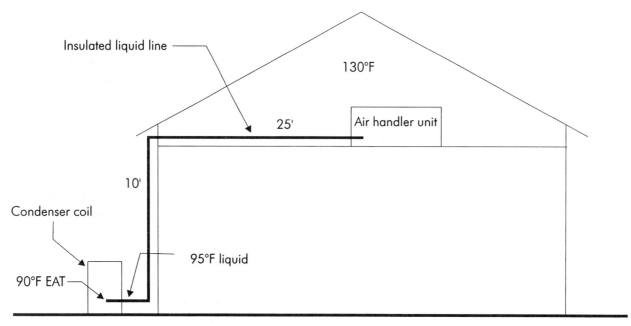

Figure 6-1. Air handler in attic

temperatures are higher than they are around most other surfaces. To reduce the effect of roof temperatures, the liquid line must be insulated and suspended at least 18 in. above the surface of the roof, Figure 6-2.

Water gas tars used as binding and waterproofing agents for roofing have a softening temperature from 104°C (219°F) for coal tar pitch to 112°C (234°F) for oil gas tar. Liquid lines on the roof may sink into the soft tar coating when the sun heats the tar to the softening temperature. The roof receives direct sun, raising its temperature, so the evaporator is receiving liquid refrigerant near the original condensing temperature. The liquid could reach the liquid saturation temperature and cause the refrigerant to vaporize before entering the metering device.

In TXV systems, the high-temperature vapor entering the evaporator will cause the TXV to open and increase the evaporator operating temperature. This reduces the split of the evaporator and greatly reduces the system capacity. In capillary tube systems, the vapor produced in the liquid line entering the capillary tubes reduces the ability of the capillary tubes to carry refrigerant. Vapor does not flow through capillary tubes as readily as liquid does. The evaporator pressure drops because of reduced refrigerant flow, and system capacity is greatly reduced.

The rule that bears repeating is **the liquid line must be insulated when exposed to higher-than-normal ambient temperatures. In addition, the line must be at least 18 in. above a flat roof.**

ELEVATION

The weight of liquid refrigerant can greatly affect system performance. A column of R-22 weighs 0.5 lb/ft of column height. When the evaporator is located higher than the condenser coil, the liquid refrigerant pressure is reduced 0.5 lb/ft of elevation. When the condensing unit is located higher than the evaporator, the weight of the refrigerant increases the liquid pressure at the metering device.

EVAPORATOR ABOVE CONDENSING UNIT

In Figure 6-3, the evaporator is located 50 ft above the condensing unit. The liquid pressure at the metering device is the condensing pressure at the condenser coil less the weight

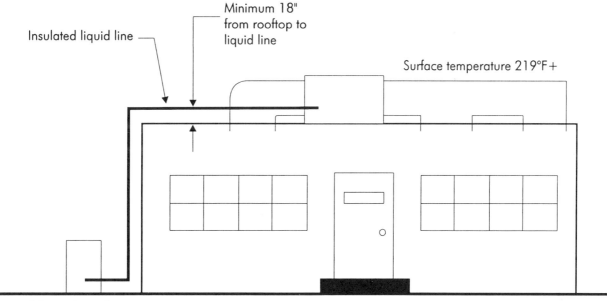

Figure 6-2. Roof installation

of the liquid. This means a 50 psig drop in liquid pressure and corresponding drop in liquid boiling point at the metering device. If the metering device is a TXV, the reduction in pressure will have the effect of forcing the valve to open to produce the superheat setting of the valve. If the metering device is a capillary tube or restrictor, the reduction in pressure will decrease the liquid refrigerant flow into the evaporator.

This reduction in pressure lowers the boiling point of the refrigerant in the evaporator, reduces the evaporator's capacity (because it produces less vapor), and lowers system capacity. The system will not handle peak afternoon loads without a greater rise in the temperature in the occupied area, and the operating cost will rise. This is a common complaint in high-rise installations.

If the high-efficiency system installed in the high-rise application is designed to have a 20°F condenser split at 90°F outdoor ambient temperature with 15°F subcooling using R-22, the liquid temperature at the condenser coil will be 95°F and 226.4 psig. With the 50-ft elevation and 50 psig reduction in the liquid pressure, the liquid pressure drops to 176.4 psig and the boiling point drops to 93°F. This boiling point is lower than the liquid temperature. Refrigerant will boil off to lower the temperature of the liquid to the boiling point before it enters the metering device. Insulation of the liquid line will not entirely cure this problem. A heat exchanger

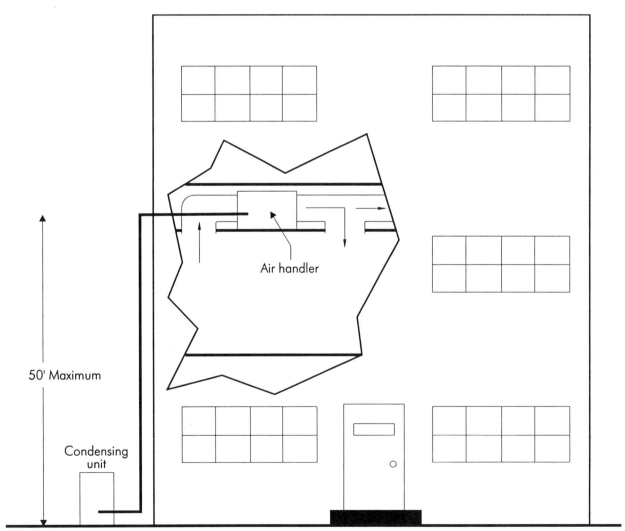

Figure 6-3. Evaporator above condensing unit

must be installed immediately ahead of the metering device to cool the refrigerant below its boiling point to restore the liquid to 100% liquid.

If the elevation pressure loss in the liquid refrigerant is combined with high ambient temperatures, the rise in the temperature of the liquid will cause the expansion to start at a lower elevation. Each 5°F rise in liquid temperature will drop the critical elevation 3 ft. The critical elevation is the point at which the boiling point drops to the sensible temperature of the refrigerant. With high-efficiency units operating at low condensing pressures, a safe elevation is under 50 ft (five stories).

Evaporator Below Condensing Unit

In Figure 6-4, the air handler is in the lower apartment, and the condensing unit is on the roof. The pressure at the inlet of the metering device equals the condensing pressure of the condensing unit plus the weight of the liquid refrigerant. This situation presents two problems: 1) liquid pressure at the metering device; and 2) oil return in the vapor line.

If the metering devices are capillary tubes or restrictors, which have no control over refrigerant flow rate, the additional pressure on the refrigerant due to the weight of the refrigerant above the condensing pressure will cause flooding of refrigerant in the evaporator. If the refrigerant charge is reduced to reduce the flooding, the system will have poor subcooling, excessive flash gas, and low system capacity. If the metering device is a TXV, the valve will close down and compensate for the higher inlet pressure; therefore, a TXV is a must in situations of this type.

Use of a TXV, however, creates an oil-return problem. The TXV does not allow the evaporator to fill with refrigerant when the system stops. If the rise in the vapor line leaving the evaporator is more than 4 ft, the oil in the system can accumulate in the evaporator if a means of keeping the evaporator clear of oil is not included. To promote good oil return and prevent oil from blocking the evaporator, oil traps must be installed in the vapor line. The distance in elevation between the traps will depend upon the velocity of the vapor in the vapor line. Vapor lines sized based on a pressure drop of 3 psig per 100 equivalent feet will carry oil up a vertical line for 10 ft. Thus, an oil trap must be installed every 10 ft.

Figure 6-5 shows the arrangement of a typical oil trap. Located in the vapor line riser, the trap consists of four 90° elbows, each of which has a flow resistance equal to 3 ft of straight pipe. Each trap installed in the riser adds the flow resistance equivalent of 12 ft of straight pipe. With the vapor line sized for a pressure drop of 3 psig per 100 ft of pipe, each trap adds 0.36 psig pressure drop in the line. The first trap is always installed at the vapor outlet of the evaporator.

Figure 6-6 shows a single oil trap arrangement on the evaporator outlet. This trap is used to catch and retain the oil traveling with the refrigerant to keep it from accumulating in the evaporator. The trap consists of three 90° elbows for an equivalent length of 9 ft. With the power element bulb of the TXV located downstream from the trap, the valve action is increased to ensure oil leaves the evaporator.

From this trap, another trap is installed in the vertical vapor line at every 10 ft of elevation. Consider a situation in which there is a 100 ft difference in elevation with the evaporator at the lower level. An evaporator outlet trap plus a trap every 10 ft to the 90 ft level will add one trap of 9 equivalent feet plus 9 traps of 12 equivalent feet. This amounts to a total pipe length of 100 actual feet plus 117 equivalent feet in the traps, or a total of 217 feet. At 3 psig per 100 ft, the pressure drop in the vapor line will be 6.5 psig. This drop in vapor pressure will reduce the capacity of the system approximately 6%.

Troubleshooting 69

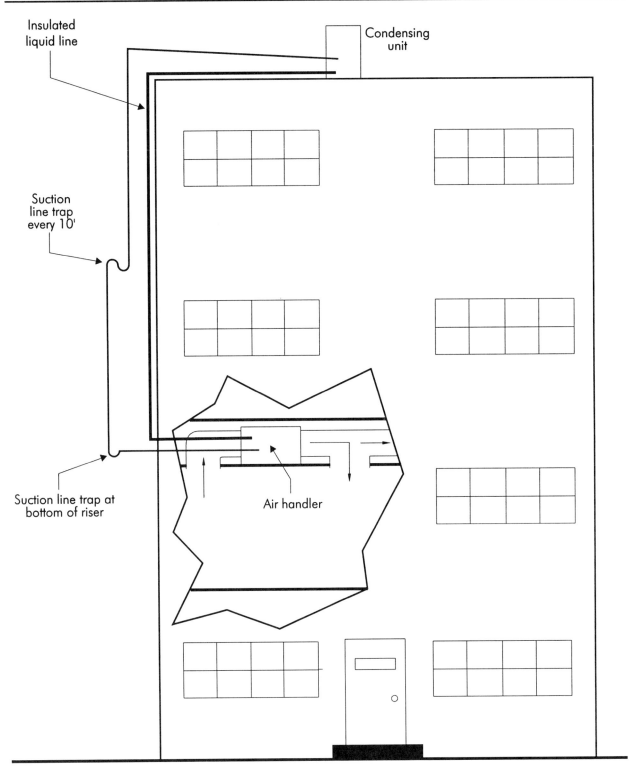

Figure 6-4. Evaporator below condensing unit

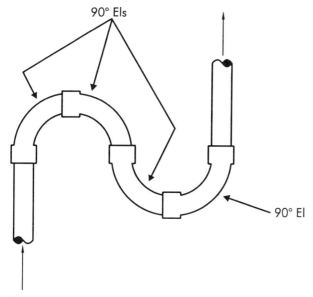

Figure 6-5. Oil trap installation

Traps are not required in systems using capillary tubes or restrictors if the condensing unit is not more than 65 ft above the evaporator. All evaporators using capillary tubes or restrictors are the bottom feed type, which practically eliminates the possibility of liquid refrigerant runout in the off cycle of the system. The refrigerant charge is also limited to what the evaporator will hold in the off cycle. Because the evaporator fills with liquid refrigerant in the off cycle, when the pressures in the systems balance, the liquid refrigerant in the evaporator will absorb heat energy to the temperature of the return air entering the evaporator.

Immediately upon startup of the system, the pressure drops in the evaporator and a high quantity of vapor is produced in all parts of the evaporator circuits. This high volume of vapor entering the vapor line will produce vapor flow velocity higher than 6000 fpm, which can lift the oil in properly sized risers up to 65 ft. This is what puts the limit on the elevation in capillary tube and restrictor systems. For installations with elevations higher than 65 ft, a TXV must be used.

VARIABLE-SPEED SYSTEMS

If the system changes the capacity of the compressor by either varying the number of cylinders used or changing the speed of the

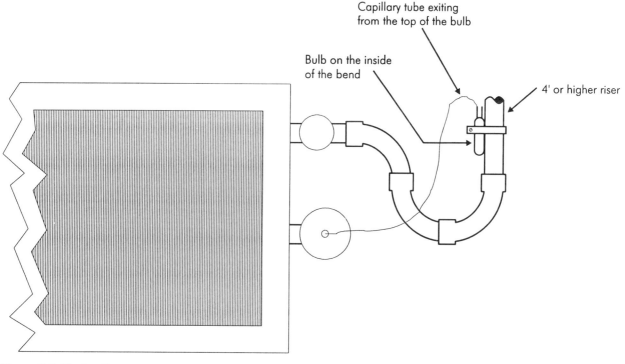

Figure 6-6. Single oil trap

Troubleshooting 71

compressor, a double-riser system is required in the vertical portion of the vapor line.

Figure 6-7 shows a double-riser arrangement. When maximum cooling is required, the system will run at full capacity and both risers will carry refrigerant vapor and oil. On minimum load, the amount of refrigerant vapor produced in the evaporator decreases. This reduces the velocity in the risers to a point at which it will not carry the oil upward. The velocity drops below the minimum of 1500 fpm. The low vapor velocity allows the oil to accumulate in the traps. The trap off the bottom of the evaporator outlet connection (A) will receive the larger portion of the oil as it travels along the bottom of the outlet connection. When this trap fills with oil, it acts as a shutoff for the secondary riser it feeds. This forces the vapor into the upper trap and into the primary riser. The vapor velocity through the primary riser is then above the 1500 fpm minimum, and the trap system returns the oil to the compressor.

If the load increases to require full capacity or if the system cycles off and on, the vapor quantity off the evaporator will produce very high velocities in the riser system at startup. The pressure difference creating the high velocity will blow the oil out of the secondary trap (A) and up the riser. This puts both risers into operation. The action of the lower (secondary) trap acts as an automatic shutoff of the secondary riser under light loads and permits full flow capacity under full loads. To function properly, the riser pipes must be sized to provide a minimum of 1500 fpm vapor velocity.

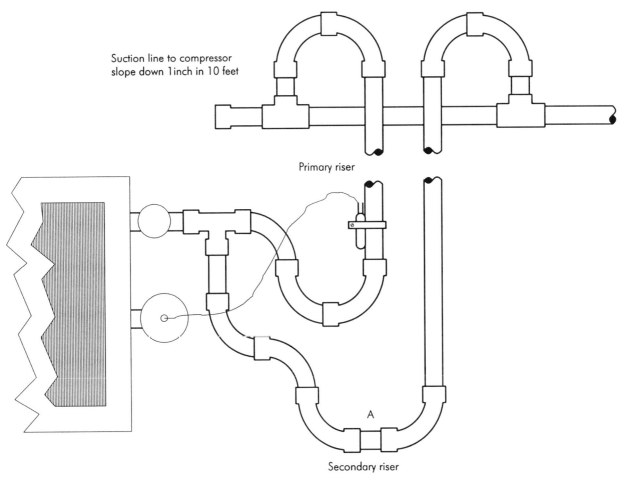

Figure 6-7. Double-riser system

LIQUID AND VAPOR LINE SIZING (R-22)

Figure 6-8 is a chart used to find vapor and liquid lines sizes for R-22 systems. Assume the system is a 3-hp air conditioning system that has two compressor speeds. The manufacturer specifies 36,000 Btuh at high speed and 18,000 Btuh at low speed. The primary riser (the riser off the trap at the end of the vapor connection on the evaporator) is sized to carry all the vapor at low speed. Both risers are sized to carry the vapor at high speed. The primary (low-speed or minimum-capacity) riser will carry the vapor produced with 18,000 Btuh or 1.5 ton load.

From the top of the chart at 1.5 tons of refrigeration, follow the vertical line down to the 40°F (rating standard) evaporator temperature line. From this point the line moves horizontally to the vertical dotted line from 1500 fpm at 120°F condenser. For proper oil return, 1500 fpm velocity and 120°F condenser are standard for air-cooled air conditioning systems. The point where these two lines meet, between 5/8" and 3/4" tubing, is the line size for the vapor (suction) line. The minimum velocity of 1500 fpm is needed to maintain good oil return; therefore, select the smaller of the two sizes — 5/8-in. od.

The capacity in Btuh of the secondary riser is the difference between the full capacity and the minimum capacity of the system. In the example, the difference is 18,000 Btuh or 1.5 tons of refrigeration. Using the same process of following the lines on the chart, the secondary riser is also 5/8 in. For this installation, two 5/8-in. risers are used to ensure proper oil return. When selecting the riser sizes in double-riser systems, the minimum capacity riser (the primary riser) is selected first.

The risers must empty into the horizontal line to the compressor to prevent oil from backing down the risers when the system cycles off. After the oil enters the horizontal vapor line, it must travel in only one direction — to the compressor. A slope of at least 1 in. for every 10 ft of the horizontal vapor line will ensure that oil will flow down.

PROBLEMS

6.1. Problems in an air conditioning system can be divided into two categories. What are they?

6.2. A system operates normally with 16°F subcooling, but it is now operating with 25°F subcooling. What is the possible cause of this change?

6.3. The system uses a TXV. As refrigerant is added to the system, the superheat decreases. True or false?

6.4. When the setting of a TXV is changed, what is the only change that occurs?

6.5. What effect does a rise in elevation of the liquid line have on the system?

6.6. What is the maximum recommended height of the evaporator above the condensing unit?

6.7. When is insulation required on the liquid line?

6.8. What effect will locating the condensing unit above the evaporator have on the operation of the system if a TXV is used?

6.9. What effect will locating the condensing unit above the evaporator have on the system if a capillary tube or restrictor is used?

6.10. Using 90° elbows, what is the equivalent length in feet of the first trap off the evaporator?

6.11. Using 90° elbows, what is the equivalent length in feet of the trap in the suction (vapor) line riser?

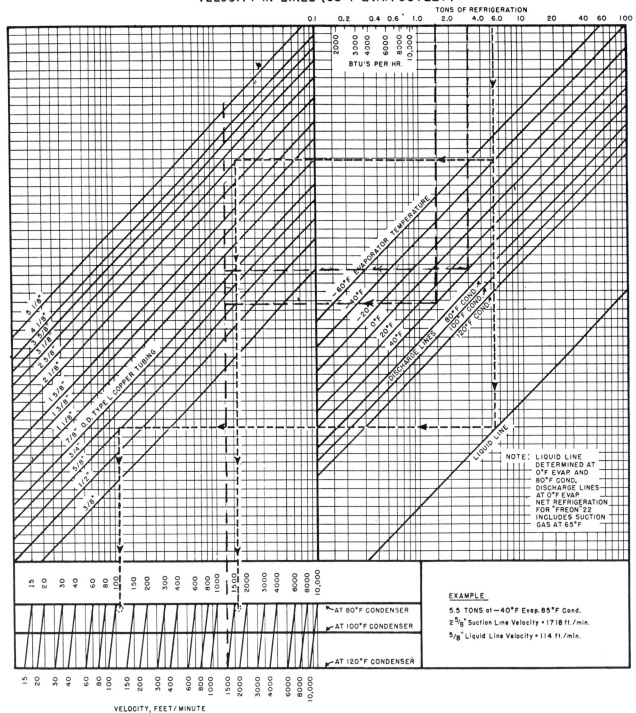

Figure 6-8. R-22 liquid and vapor line sizes (Courtesy, The Trane Company)

Glossary of Technical Terms

A

ACCUMULATOR: A storage vessel located in the suction line ahead of the compressor. Used to limit liquid refrigerant return to the compressor.

AIR CONDITIONER: A piece of equipment used to provide control of temperature, humidity, air cleanliness, and air distribution in a specific area.

AIR CONDITIONING: The simultaneous control of temperature, humidity, cleanliness, and air distribution for human comfort.

AIR CHANGES: A method of expressing the amount of air leakage into or out of a building or room interior or the number of building or room volumes exchanged per hour.

AIR-COOLED CONDENSER: A piece of equipment that transfers heat picked up by the air conditioning system into the surrounding atmosphere.

AIR COOLER: A device used to lower the temperature of air passing through it.

AIR DIFFUSER: An air distribution outlet designed to direct airflow in desired directions.

AIR HANDLER: A device used to move air. Contains a blower assembly for active air movement.

AIR MOVEMENT, ACTIVE: Air moved by means of a mechanical device such as a fan or blower.

AIR MOVEMENT, PASSIVE: Air moved by the difference in the weight of the air due to the difference in temperature of the air.

AIR RECIRCULATION: The movement of air directly from a unit's discharge opening to the unit intake opening. This seriously affects the operation of the unit by creating adverse temperatures.

AIR RETURN: Air returned from a conditioned space.

AIR, SATURATED: A sample of air that contains the maximum amount of water vapor. The percentage of relative humidity in the air is 100%.

AIR, SPECIFIC HEAT OF: The amount of heat needed to change the temperature of one pound of air (at the existing temperature) 1°F as compared to the amount for an equal amount of standard air.

AIR, STANDARD: Air with a density of 0.075 lb/ft^3 at 70°F and 29.92 in. Hg atmospheric pressure.

ALTERNATING CURRENT (ac): Electrical current in amperes, that reverses direction. As the voltage or electrical pressure varies — rising to maximum, falling to zero — and reverses direction — rising to maximum and falling to zero — the current in the circuit follows the change. Sixty hertz or cycle frequency refers to these 360-degree reversals.

AMBIENT TEMPERATURE: The temperature of the air that surrounds an object.

AMMETER: An electrical meter used to measure electric current in amperes.

AMPACITY: A term used to designate the amperage carrying capacity of wires or conductors.

AMPERE: A unit of electrical energy quantity.

ANEMOMETER: An instrument used to measure the flow rate of air.

B

BACK PRESSURE: Pressure in the low side of the system. Also called low-side pressure or suction pressure.

BAROMETER: An instrument for measuring atmospheric pressure, otherwise known as barometric pressure. The predominant calibration is in inches of mercury (Hg) in a vertical column.

BLOWER: A device using a centrifugal fan that provides pressure to move air.

BOILING POINT: The temperature at which a liquid boils, changing from a liquid to a vapor at a given pressure.

BRITISH THERMAL UNIT (Btu): The quantity of heat energy added to or removed from one pound of water to change the temperature 1°F.

C

CALCULATOR, GAS-SERVICE: A slide rule calculator used to determine the orifice size for a vapor-burning fossil fuel heating unit.

CAPACITY, NET COOLING: The cooling capacity of an air conditioning unit or heat pump in the cooling mode. This is made up of the sensible and latent heat picked up by the evaporator.

CAPILLARY TUBE: A fixed, restriction-type metering device. Usually consists of lengths of small inside diameter tubing. The flow restriction provides the necessary reduction in pressure and boiling point of the liquid refrigerant before it enters the evaporator.

CELSIUS: Temperature scale used in the metric system. Water freezes at 0°C and boils at 100°C.

CHARGE, REFRIGERANT: The quantity of refrigerant in the system.

CHARGING: The process of putting refrigerant into the system.

COIL, INSIDE: The coil located in the inside portion of the heat pump system. Performs as an evaporator coil in the cooling mode and as a condenser in the heating mode.

COIL, OUTSIDE: The coil located in the outside portion of an air-to-air heat pump system. Performs as an evaporator in the heating mode and as a condenser in the cooling mode.

COMFORT CHART: Chart used to show the dry bulb temperature and humidity conditions for human comfort conditions.

COMFORT ZONE: The area on a psychrometric chart that shows the conditions of temperature and humidity in which humans encounter some degree of comfort.

COMPRESSOR: The pump of the air conditioning system that intakes refrigerant vapor and raises its temperature and pressure to the point where it can be condensed for re-use.

COMPRESSOR, HERMETIC: An assembly in which the motor and compressor are one assembly inside a welded-together housing.

COMPRESSION GAUGE: (See GAUGE, HIGH-PRESSURE).

CONDENSATE: Water that condenses out of air passing over an evaporator.

CONDENSATE DRAIN TRAP: A pipe arrangement that provides a water seal in the drain line to prevent air from flowing through the drain line.

CONDENSATE PAN: A pan located under an evaporator to catch the condensate of the evoporator and carry it to the drain line.

CONDENSATION: Liquid that forms when a vapor is cooled below its condensing temperature.

CONDENSE: The action of changing a saturated vapor to a saturated liquid.

CONDENSER: The part of the air conditioning system in which compressed refrigerant vapor is cooled until it becomes a liquid.

CONDENSER, AIR-COOLED: A heat exchanger that transfers heat energy from the refrigerant vapor to the air.

CONDENSER FAN: A device used to force air through an air-cooled condenser.

CONDENSER, LIQUID-COOLED: A heat exchanger that transfers heat energy from the refrigerant vapor to the liquid heat sink.

CONDENSING TEMPERATURE: The temperature at which a vapor changes to a liquid at a given pressure.

CONDENSING UNIT: The portion of the air conditioning system that intakes low-temperature/pressure refrigerant and converts it to a liquid suitable for the absorption of more heat; the compressor, condenser, receiver, and their controls. Commonly called the high side of the system.

COOLING MODE: The operating phase of a system that is removing heat and/or moisture from a conditioned area.

CONTROL, REFRIGERANT: A device used to provide the necessary pressure reduction on the liquid refrigerant to obtain its proper boiling point in the evaporator.

CONTROL, TEMPERATURE: A device that uses changes in air temperature to operate contacts in an electrical circuit.

CONVECTION: Transfer of heat energy by means of a flow of liquid or vapor.

CONVECTION, ACTIVE: Transfer of heat energy by the forced movement of a liquid or vapor.

CONVECTION, PASSIVE: Circulation of a liquid or vapor due to differences in weight of the liquid or due to differences in temperature of the vapor.

CYLINDER, REFRIGERANT: A cylinder in which refrigerant is purchased or recovered from a system and from which refrigerant is dispersed. It is color-coded according to the type of refrigerant it contains.

D

DEHYDRATE: To remove water in all forms from a material or system.

DEHUMIDIFY: To remove water vapor from the atmosphere.

DENSITY: Closeness of texture or consistency.

DESIGN TEMPERATURE DIFFERENCE: The difference between the design indoor and outdoor temperatures. Design temperature is the assumed outdoor temperature when figuring heat loss or gain.

DETECTOR, LEAK: A test instrument used to detect and locate refrigerant leaks.

DEW POINT: Temperature at which water vapor begins to condense and fall out of air. Air at the dew point, 100% relative humidity, is referred to as saturated vapor.

DICHLORODIFLUOROMETHANE: A refrigerant also called R-12. Its chemical formula is CCl_2F, its cylinder color code is white, and its boiling point is 29.92 in. Hg, standard atmospheric pressure, or -21.6°F.

DISCHARGE LINE: The piping between the compressor discharge port and condenser. Also called the hot gas line.

DRAIN PAN: See CONDENSATE PAN.

DRIER: A substance or device used to prevent moisture from circulating through a refrigeration system.

DRY BULB TEMPERATURE: The actual (physical) temperature of a substance. Usually refers to air.

DRY BULB THERMOMETER: An instrument with a dry sensing element that measures the physical temperature of a substance.

DUCT: A round or rectangular metal or fiberglass pipe that carries air between the air conditioning unit or heat pump and the conditioned area.

E

EFFECTIVE AREA: The actual opening in a grille or register through which the air can pass. The effective area is the gross (overall) area minus the area of the deflector bars or vanes.

EFFECTIVE TEMPERATURE: The effect of temperature, humidity, and air movement on a human being.

EFFECTIVE TEMPERATURE DIFFERENCE: The apparent temperature sensed by people in a conditioned space. It is an index, or a measure, that combines dry bulb temperature, humidity, and air motion.

ELECTRIC HEATING: A heating system that uses electric resistance elements as the heat energy source.

ELECTRIC HEATING ELEMENT: A unit consisting of resistance wire, insulated supports, and connection terminals for connecting the resistance wire to a source of electrical energy.

ELECTRONIC LEAK DETECTOR: An electronic instrument that measures the changes in electron flow across an electrical gap, which indicates the presence of refrigerant vapor molecules.

ELECTRONICS: Field of science dealing with electronic devices and their application.

ENERGY EFFICIENCY RATIO (EER): The comparison of the heat-transferring ability of an air conditioning system and the electrical energy used, expressed in Btu/watt.

EVACUATE: See DEHYDRATE.

EVAPORATION: The term used to describe the changing of a liquid to a vapor by the addition of heat energy.

EVAPORATOR: The coil in which refrigerant absorbs heat and evaporates.

EXFILTRATION: The outward flow of air from an occupied area through openings in the building structure.

EXHAUST OPENING: An opening through which air is removed from an area.

EXTERNAL EQUALIZER: A pressure connection in the area below the diaphragm of a TXV and the suction outlet of an evaporator.

F

FAHRENHEIT: Temperature scale on which water freezes at 32°F and boils at 212°F.

FAN: A radial or axial flow device using blades for moving air.

FAN, CENTRIFUGAL: A fan rotor or a wheel within a blower scroll, which includes either a direct drive motor or pulley-and-belt drive combination.

FAN, PROPELLER: A blade-type wheel within a mounting ring that has either a direct-drive motor or pulley-and-belt combination.

FILTER: A device for removing foreign particles from vapor or liquid.

FIRE-EFFICIENCY FINDER/STACK LOSS RULE: A slide rule that uses the CO_2 percentage and temperature of a vapor to determine the efficiency of a fossil fuel-powered heating unit.

FLASH GAS: The vapor resulting from the instantaneous evaporation of refrigerant as it leaves a metering device. The refrigerant is thus cooled to the temperature corresponding to the pressure in the evaporator.

FLOWHOOD: An instrument that measures air quantity (cfm) coming out of or going into a duct system and directs the air over a sensing element.

FLOW METER: An instrument used to measure the volume of a liquid flowing through a pipe or tube in quantity per time (gallons per minute, pounds per hour, etc.).

FLUID: A substance in a liquid or vapor state that moves and changes position without separation of the material.

FLUID FLOW: The movement of a liquid by a mechanically created pressure difference or a difference in fluid density created by a temperature difference.

FORCED CONVECTION: Movement of a liquid or vapor by a mechanical means.

FREE AREA: The area of the opening in a grille or register through which air passes.

FREEZING: The conversion of a liquid to a solid.

FREEZING POINT: The temperature at which a liquid becomes a solid upon the removal of latent heat.

FYRITE INDICATOR, CO_2: An instrument used to measure the percentage of CO_2 in a vapor sample.

FYRITE INDICATOR, O_2: An instrument used to determine the percentage of oxygen (O_2) in a vapor sample.

G

GAS: A substance in the vapor state.

GAS SUPPLY METER: A device that measures and records on indicator dials the amount of gas in cubic feet that flows through the meter.

GAS VALVE: A device that controls the flow of gas or vapor.

GAUGE, COMPOUND: An instrument that measures pressures above and below atmospheric pressure.

GAUGE, GAS MANIFOLD PRESSURE: A gauge for reading the gas manifold pressure of a vapor-burning heating unit. A more convenient substitute for a U-tube manometer. Measures in increments of inches of water.

GAUGE, HIGH-PRESSURE: An instrument used to measure pressures above atmospheric pressure in the 0 to 500 psig range.

GAUGE, LOW-PRESSURE: An instrument used to measure pressures above atmospheric pressure up to 50 psig.

GAUGE, MANIFOLD: A device that contains a combination of gauges and control valves to control the flow of vapors or liquids through the device.

GAUGE, STANDARD: An instrument designed to measure pressures above atmospheric.

GAUGE, VACUUM: An instrument used to measure pressures below atmospheric.

GRAVITY, SPECIFIC: The specific gravity of a solid or liquid, which is the ratio of the mass of the material to the mass of an equal volume of water at standard temperature (70°F). The specific gravity of a vapor is usually expressed in terms of dry air at the same temperature and pressure of the vapor.

GRILLE: An ornamental or louvered opening through which air leaves a duct system or area.

GROSS HEATING CAPACITY: The gross capacity of a heat pump in the heating mode, which is the total amount of heat energy transferred to the inside air or material. This is made up of the heat energy picked up from a heat source (air or liquid) plus the heat energy equivalent of the electrical energy required to operate the motor-compressor assembly and outdoor fan.

H

HALIDE REFRIGERANTS: Refrigerants that contain halogen chemicals such as chlorine and fluorine in their molecular structure.

HALIDE TORCH: A leak detection device that produces a reaction in a flame when it detects halogen chemicals in the air.

HEAD PRESSURE: The pressure in the condensing or high side of the air conditioning system.

HEAD PRESSURE CONTROL: A pressure-operated safety control that opens an electrical circuit if the pressure exceeds the cut-out point of the control.

HEAT: The energy that affects the molecular activity of a substance. This is reflected in the temperature of the substance. Addition of heat energy increases the molecular activity and temperature. Removal of heat energy lowers the molecular activity and temperature.

HEAT, LATENT: Heat energy used to change the state of a substance without a change in the temperature of the substance. A substance changes from solid to liquid to by the addition of latent heat and from vapor to liquid to solid by removal of latent heat.

HEAT, SENSIBLE: The heat energy used to change the temperature of a substance without a change in state.

HEAT, SPECIFIC: The heat required to change the temperature of a material. Heat quantity is measured per pound of material per °F.

HEAT, TOTAL: The sum of both the sensible and latent heat energy in air. Expressed as Btu/lb of air. The term used to express total heat in air is enthalpy.

HEAT EXCHANGER: A device used to transfer heat energy from an area with a higher temperature to one with a lower temperature. Evaporators and condensers are examples.

HEATING COIL: A heat-transfer device designed to add heat to a liquid or vapor.

HEATING CONTROL: A device that controls the operation of a heating unit to maintain a set temperature in a material or vapor.

HEATING MODE: The operating phase of a system that adds heat to an occupied area.

HEATING UNIT, ELECTRICAL: A device containing one or more electrical resistance elements, electrical connections, and safety and control devices in a frame or casing.

HEAT OF FUSION: The heat energy needed to accomplish the change in state of a material between a liquid and solid. Heat must be added for a change from solid to liquid and removed for a change from liquid to solid.

HEAT LOAD: The amount of heat energy, measured in Btuh, that is required to maintain a given temperature in a conditioned area at design conditions both inside and outside the conditioned area.

HEAT PUMP: A mechanical compression cycle refrigeration system that can be reversed to either heat or cool a conditioned area.

HEAT PUMP, AIR-TO-AIR: A device that transfers heat between two different air quantities in either direction on demand.

HEAT PUMP, AIR-TO-LIQUID: A device that transfers heat from an air source to a liquid by means of a refrigeration system. Units in this category are non-reversible.

HEAT PUMP, LIQUID-TO-AIR: A device that transfers heat between liquid and air in either direction on demand.

HEAT PUMP, LIQUID-TO-LIQUID: A device that transfers heat between two liquids in either direction on demand.

HEAT PUMP, WATER HEATER: An air-to-liquid refrigeration system used to heat domestic water with air as the heat source.

HEAT RECOVERY UNIT: A heat exchanger in the hot gas line between the compressor discharge and the condenser. Used to heat water by removing the superheat from the compressor vapor.

HEAT SINK: A place or material into which heat energy is placed.

HEAT SOURCE: A place or material from which heat energy is obtained.

HEAT TRANSFER: The movement of heat energy from one body or substance to another. Heat may be transferred by any combination of or by all three methods of radiation, conduction, and convection.

HEATING VALUE: The amount of heat energy released by the burning of a fuel or operation of an electric element. It is usually expressed in Btu/kW for electricity; Btu/ft^3 for vapor; Btu/gph for liquids; and Btu/lb for solids.

HERMETIC MOTOR: A motor completely sealed in a welded case.

HERMETIC SYSTEM: An air conditioning system that uses the hermetic motor-compressor assembly in a welded-together system.

HIGH SIDE: Any part of an air conditioning system under high pressure; that section of an air conditioning system starting at the compressor discharge and extending to the metering device.

HIGH VACUUM PUMP: A pump designed to create a vacuum on the intake side of the pump of less than 1000 microns pressure.

HUMIDITY: Moisture in air.

HYGROMETER: An instrument used to measure the percentage of moisture (relative humidity) in air.

I

INCH OF MERCURY COLUMN: A unit of pressure measurement. One inch of mercury column is equal to a pressure of 0.491 psig.

INCH OF WATER COLUMN: A unit of measurement for low gas pressures; approximately 1/28 of one psi.

K

KILOWATT: A unit of electrical energy equal to 1000 watts.

L

LIQUID LINE: The refrigerant line or tube that carries liquid refrigerant from the condenser to the metering device.

LOAD: The amount of heat that the refrigerant system is required to remove or supply at design conditions. The amount of electrical energy expected to be connected to an electrical power supply.

LOW SIDE: The portion of the air conditioning system that operates at the evaporator pressure. Consists of the metering device, evaporator, and vapor line.

M

MANIFOLD, SERVICE: See GAUGE, MANIFOLD.

MANOMETER: An instrument used to measure the pressure of gases or vapors. The gas pressure is balanced against a column of liquid such as water or mercury in a U-shaped tube open to atmospheric pressure.

MANOMETER, INCLINED: A manometer on which the liquid tube is inclined from horizontal to produce wider, more accurate readings over a smaller range of pressures.

METER: A metric length or distance, equal to 39.37".

METER, FLOW: See FLOW METER.

METERING DEVICE: The device used to reduce the pressure on the liquid refrigerant and thus the boiling point before the refrigerant enters the evaporator.

METRIC SYSTEM: A system of weights and measures based on multiples of ten; a decimal system; also known as SI system.

MICRON: A metric unit that measures length and is about one millionth (1/1,000,000) of a meter. There are 25,400 microns in one inch.

MICRON GAUGE: An instrument used to measure pressures below atmospheric (vacuum) or below 1 in. Hg.

N

NATURAL CONVECTION: Movement of air or liquid caused by differences in weight of the material as a result of the difference in temperatures of the material.

NITROGEN: An inert gas used in portions of tubing and/or parts that are subjected to high heat during the assembly process. The nitrogen replaces the oxygen in the air in the tube or part and reduces the formation of scale (copper oxide).

NOMINAL-SIZE TUBING: Tube measurement that has an outside diameter the same size as the inside diameter of iron pipe of the same given size.

O

OCCUPIED AREA: The area that is conditioned by a heat pump, air conditioner, or heating unit.

OFF MODE (CYCLE): That part of the operating mode when the system has been shut down by the controls.

P

PIPE, FRICTION LOSS: The flow resistance of liquid through a pipe. Dependent on pipe size and flow quantity. Expressed in feet of head per 100 ft of pipe.

PITCH: Pipe or tube slope in the direction of flow to cause gravity to enhance the flow of material through the pipe or tube.

PITOT TUBE: A tube used with a manometer to measure air velocities and pressures.

PLENUM: A cube-shaped box or duct on the supply and return sides of an air handling unit that connects the supply and return duct system to the unit.

POINT OF VAPORIZATION: The location of the position in the evaporator circuits where the last bit of liquid refrigerant is vaporized.

PRESSURE: An energy impact on a unit area. Also force or thrust against a surface.

PRESSURE DROP: The pressure difference between two locations in a circuit that is the result of the flow resistance in the circuit.

PRESSURE, STATIC: The pressure in the system when the compressor is idle.

PRESSURE, SUCTION: The refrigerant pressure in the suction line, at the compressor inlet.

PSYCHROMETER: An instrument used to measure the dry bulb and wet bulb temperatures of air.

PSYCHROMETRIC CHART: A chart that graphs the relationship of the temperature, pressure, and moisture content of air.

PSYCHROMETRIC MEASUREMENT: Using a psychrometer to measure the dry bulb and wet bulb temperatures of air and a psychrometric chart to determine the characteristics of the air.

PURGING: Releasing compressed gas into the atmosphere through ports or parts of a refrigeration system to remove excess refrigerant or contaminants. This practice is no longer legal.

R

REFRIGERANT: A liquid material used in a refrigeration system to absorb heat energy in the vaporization (liquid to vapor) process and reject heat energy in the condensing (vapor to liquid) process.

REFRIGERANT CHARGE: The quantity of refrigerant that a system requires in order to operate properly.

REFRIGERANT MIGRATION: The transfer of refrigerant in a system from a high-temperature location to a low-temperature location. The transfer results from the pressure difference created by the temperature difference.

REFRIGERATING EFFECT: The amount of heat energy in Btuh or calories per hour that the system is capable of transferring under the given circumstances.

REFRIGERATION: The process of transferring heat energy from one place to another by the change in state of a liquid refrigerant.

REGISTER: A device that combines a grille and damper assembly to control the quantity and direction of airflow.

REHEAT: The addition of heat energy into air after it has been cooled.

RELATIVE HUMIDITY: The percentage of moisture in air compared to the amount of moisture in saturated air at the same pressure and temperature conditions.

REMOTE (SPLIT) SYSTEM: A refrigeration system, the high side and low side of which are in different locations and are connected by refrigerant lines.

S

SATURATION: A condition existing when a substance contains the maximum amount of another substance for that pressure and temperature; for example, moisture in air, antifreeze in water, etc.

SENSIBLE HEAT: Heat energy added to or removed from a material that causes a change in physical temperature without a change in state of the material.

SLING PSYCHROMETER: An instrument rotated to move air over its sensing bulbs, which measure the dry bulb and wet bulb temperatures of the air.

SPECIFIC GRAVITY: The weight of a quantity of a liquid at a given temperature and pressure as compared to a like quantity of water at the same temperature and pressure. The specific gravity of water is 1.000.

SPECIFIC VOLUME: The volume of a given weight of a substance at a given temperature.

SPLIT, CONDENSER: The difference in temperature between the entering air temperature and the condensing temperature of the refrigerant vapor in the condenser.

SPLIT, EVAPORATOR: The difference in temperature between the average temperature of the air flowing through the evaporator and the average boiling point of the refrigerant in the evaporator.

SPLIT SYSTEM: See REMOTE (SPLIT) SYSTEM.

SQUIRREL CAGE FAN: The wheel of a centrifugal fan, so called due to its resemblance to rotary exercise wheels.

STANDARD ATMOSPHERIC CONDITIONS: Used as the basis for calculation and testing. It is air at 14.7 psia (pounds per square inch, absolute) and 70°F.

STATIC PRESSURE: The pressure exerted against the inside surfaces of a container or duct. Sometimes called bursting pressure.

STRATIFICATION OF AIR: The condition in which the movement of air is less than 50 fpm.

SUBCOOLING: The removal of heat from a liquid that is below its condensation point.

SUCTION LINE: Tubing or pipe used to carry refrigerant vapor from the evaporator to the compressor in air conditioning and refrigeration systems or from the reversing valve to the compressor in dual-function heat pumps.

SUPERHEAT: The heat added to a vapor to raise its temperature above its boiling point.

T

TEMPERATURE: The degree of heat energy in a material measured by a thermometer.

TEMPERATURE-HUMIDITY INDEX: The temperature and relative humidity of a sample of air compared to standard conditions.

TEST DIAL: One of the recording dials on a gas meter that indicates a quantity of gas per revolution of the dial.

THERMAL BALANCE: When the heat-absorbing ability and the heat-rejecting ability of the air conditioning system are stabilized and the pressures and temperatures are constant.

THERMOCOUPLE: A device that develops an electrical pressure (voltage) due to temperature differences across hot and cold junctions of two dissimilar metals welded together.

THERMOCOUPLE THERMOMETER: An electrical instrument that uses the amount of voltage generated by a thermocouple to indicate temperature.

THERMOMETER: An instrument used to measure the temperature of materials.

THERMOMETER, DIGITAL: A thermometer that uses solid-state circuitry and direct-reading digital numbers to display temperature. Easier to read and more accurate than ordinary column or dial thermometers.

THERMOMETER, STACK: A thermometer with a range of 200° to 1000°F used to measure the temperature of flue products from fossil fuel-powered heating units.

THERMOSTATIC CONTROL: A device that controls the operation of equipment according to the temperature surrounding the control.

THERMOSTATIC EXPANSION VALVE: A metering device that varies the refrigerant flow into the evaporator to maintain a constant superheat in the vapor at the evaporator. The pressure in the evaporator connected under the diaphragm of the valve and the temperature of the feeler bulb (power element) of the valve fastened to the outlet of the evaporator are the controlling factors.

TON OF REFRIGERATION: The refrigerating effect that is equal to the melting of one ton (2000 lb) of ice in a 24-hour period. The equation is as follows:

(144 Btu/lb)(2000 lb) = 288,000 Btu/24 hr

288,000 Btu/24 hr ÷ 24 hr = 12,000 Btuh

TOTAL PRESSURE: The sum of the static pressure and velocity pressure at the point of measurement in a duct.

TUBING, PRECHARGED: A refrigerant line that is fitted with quick-couple type sealed fittings at each end. The line has been evacuated at the factory and filled with the same refrigerant used in the system.

V

VACUUM: Reduction in pressure below atmospheric pressure.

VACUUM PUMP: A high-efficiency vapor pump used for creating a deep vacuum in air conditioning systems for testing and/or dehydration.

VAPOR: A fluid in gaseous form.

VAPOR LINE: Found in dual-action heat pumps. It is the suction line in the cooling mode and the hot gas line in the heating mode.

VAPOR, SATURATED: A vapor whose temperature has been reduced to the point of condensation, but condensation has not started.

VELOCIMETER: An instrument used to measure air velocities on a direct reading airspeed indication dial.

VOLTMETER: An instrument used to read electrical pressure (voltage) in an electrical circuit.

VOLUME, SPECIFIC: The volume of a substance per unit of mass at a given temperature. The opposite of specific density. Vapor is measured in cubic feet per pound. Liquid is measured in cubic feet per gallon or pound.

Abbreviations

A

a/c — air conditioning

ACCA — Air Conditioning Contractors of America

Amp (A) — Ampere, a unit of electricity

ANSU — actual number of scale units — difference between specific volume point on the psychrometric chart at conditions of air sample and larger specific volume curve

B

Btu — British thermal unit — the amount of heat energy added or removed to change the temperature of one pound of water one degree Fahrenheit

Btu/ft^3 — Btu per cubic foot

Btuh — Btu per hour

Btu/lb — Btu per pound

Btu/W — heat equivalent of 1 watt of electrical energy (3.413 Btu)

C

°C — temperature measured on the Celsius scale

cfm — cubic feet of air per minute

CO_2 — carbon dioxide

D

db — dry bulb temperature

D — delta, Greek symbol for difference

DHt — difference in total heat content or enthalpy

DP — difference in pressure

DT (or TD) — difference in temperature

dp — dew point

DX — direct expansion

E

EAT — entering air temperature

ESP — external static pressure

et — effective temperature

F

°F — temperature measured on the Fahrenheit scale

fpm — feet per minute (velocity)

' or ft — foot, (measurement)

ft^2 — square feet

ft^3 — cubic feet

ft^3/min — cubic feet per minute

ft^3/hr — cubic feet per hour

G

gcc — gross cooling capacity

gph — gallons per hour

gpm — gallons per minute

gr/lb — grains of moisture per pound of air

H

Hl — latent heat

h/p — heat pump

Hs — sensible heat

hr — hour (time)

Ht — enthalpy — the total heat quantity of air. The sum of the sensible and latent heat in the air at the particular temperature and relative humidity.

I

" or in. — inch (length)

in^2 — square inches

in. Hg — inches of mercury (pressure on the barometric scale)

L

LAT — leaving air temperature

LP — liquefied petroleum gas (propane)

M

mg — manufactured gas

mh — motor heat

min — minute (time)

N

ncc — net cooling capacity

ng — natural gas

NSU — number of scale units — number of scale units at the specific volume of the air sample on the psychrometric chart

O

oz — ounce (weight)

P

% — percent

π — pi — 3.416 — ratio of the circumference of a circle to its diameter

PF — power factor

lb/hr — pounds per hour

lb/min — pounds per minute

Ps — static air pressure

psia — pounds per square inch, absolute

psig — pounds per square inch measured on a gauge that uses the atmospheric pressure as its base or zero-pressure point

Pt — total air pressure

R

RH — relative humidity

r^2 — radius of a circle squared

S

sec — second (time)

SH — specific heat — heat quantity needed to change a pound of air one degree Fahrenheit. Heat quantity needed to change the temperature of a liquid one degree Fahrenheit compared to an equal quantity of water.

T

TD — temperature difference

TSU — total scale units — total scale units between the cubic feet per pound lines on the psychrometric chart

TXV — thermostatic expansion valve

V

V — volts

Vp — velocity pressure

Vs — specific volume

W

W — watt of electrical power

wc — water column (pressure)

wb — wet bulb temperature

Answers

CHAPTER 1

1.1. The heat energy picked up in the evaporator plus the motor heat equal the heat energy off the condenser. Net capacity plus motor heat equal gross capacity. The balance between the net capacity plus the motor heat and the gross capacity is the thermal balance of the system.

1.2. Enthalpy is the sum of total sensible and latent heat energy in air.

1.3. Relative humidity is the quantity of moisture in air compared to the amount the air could hold when saturated, at a given temperature. It is expressed as a percentage.

1.4. Psychrometry is the science of the characteristics of air, regarding heat content, moisture content, and specific volume.

1.5. Dry bulb (db) temperature measures with a dry bulb thermometer the heat intensity of air temperature only, not humidity.

1.6. Wet bulb (wb) temperature measures with a wet bulb thermometer, the bulb of which is covered with a water-soaked wick. The lowering of temperature that results from the evaporation of water around the bulb indicates the air's relative humidity.

1.7. Dew point is the temperature at which vapor starts to condense so that it becomes a liquid.

1.8. The temperature at which moisture starts to condense out of air — 100% relative humidity.

1.9. Specific humidity is the actual weight of water vapor expressed in grains of water per pound of dry air. It also may be expressed as pounds of water per pound of dry air.

1.10. Sensible heat is the heat energy used to change the temperature of a substance without a change in state.

1.11. Latent heat is the heat energy that changes the state of a substance without a change in the temperature.

1.12. Specific volume (Vs) is the number of cubic feet (ft^3) occupied by one pound of the mixture of dry air and water vapor.

1.13. 1-C, 2-F, 3-A, 4-E, 5-H, 6-J, 7-B, 8-G, 9-D, 10-I.

CHAPTER 2

2.1. 1.08 is a constant used to convert cubic feet per minute to cubic feet per pound and from pounds per minute to pounds per hour. It also contains the quantity of heat energy needed to change the temperature of each pound of air.

2.2. The efficiency of a resistance-type heater element is 100%. The amount of heat energy equivalent to the electrical energy going into the element will result in the same amount of heat energy coming out of the element into the air. Therefore, to measure the amount of heat energy coming from the element, just determine the watts of electrical energy the element is using. The watts of electrical energy in the element per hour multiplied by 3.413 Btu/watt is used to determine the heat energy in Btuh:

Btuh = (watt/hr)(3.413 Btu/watt)

There are 1000 watts per kilowatt; therefore:

(4.0 kWh)(1000 watts per kW)(3.413 Btu per watt) = 13,652 Btuh output

2.3. The output of the heating unit is determined from the input in Btuh times the efficiency of the unit:

(120,000 Btuh)(.80 efficiency) = 96,000 Btuh output

2.4. The Btuh input to the oil-fired heating unit is based on 140,000 Btu/gallon of No. 2 fuel oil. If the nozzle used is a .85 gph, the Btuh input would be 140,000 Btu/gal at 100 psig nozzle pressure multiplied by .85 gph, which is 119,000 Btuh. The output of the heating unit is determined from the input in Btuh multiplied by the efficiency of the unit:

(119,000 Btuh)(.80 efficiency) = 95,200 Btuh output

2.5. $\dfrac{60 \text{ sec/min}}{39 \text{ sec/revolution}}$ = 1.54 revolutions of the dial per minute

1.54 revolutions per minute = 1.53 ft^3 of gas at 1050 Btu/ft^3 or:

(1.53 ft^3)(1050 Btu/ft^3) = 1606.50 Btu/min

(1606.50 Btu/min)(60) = 96,390 Btuh

(96,390 Btuh)(.80 efficiency) = 77,112 Btuh output

2.6. The unit input must first be reduced 4% per 1000 ft above sea level:

$\left(\dfrac{.04}{1000 \text{ ft}}\right)$ (8000 ft) = 32%

1.00 - 0.32 = 0.68 input

(80,000 Btuh)(0.68) = 54,400 Btuh input

2.7. $\dfrac{119,000 \text{ Btuh input}}{140,000 \text{ Btu/gallon}}$ = 0.85 gph nozzle

2.8. One half of 18 in. diameter = 9 in. radius

(9 in)^2 = 81 in^2

(81 in^2)(3.416) = 276.69 in^2

$\dfrac{276.69 \text{ in}^2}{144 \text{ in}^2/\text{ft}^2}$ = 1.92 ft^2

(1.92 ft^2)(860 fpm velocity) = 1651 cfm

2.9. (14 in)(6 in) = 84 in²

$$\frac{84 \text{ in}^2}{144 \text{ in}^2/\text{ft}^2} = 0.58 \text{ ft}^2$$

(0.58 ft²)(0.65 free area) = 0.37 ft² free area

(0.37 ft² free area)(480 fpm) = 177.60 cfm

2.10. E. -10% to +0%

CHAPTER 3

3.1. Superheat always refers to a vapor. A superheated vapor is any vapor that is above its saturation temperature for a given pressure.

3.2. Superheat is sensible heat.

3.3. The amount of superheat is the difference in temperature between the refrigerant's physical temperature and the evaporating temperature, or boiling point.

3.4. When using a capillary tube, the amount of superheat in the evaporator is determined by the evaporator air inlet wet bulb temperature and the dry bulb temperature of the air at the condenser inlet.

3.5. The maximum amount of refrigerant that should be added at one time is 4 oz.

3.6. Subcooling is the removal of heat from a liquid that is below its condensation point.

3.7. Subcooling reduces the amount of flash gas formed in the evaporator to produce the highest operating efficiency in the system.

3.8. The condenser split is the temperature difference between the condensing temperature and the ambient temperature.

3.9. False. Condenser splits vary widely, especially with high-efficiency systems.

3.10. False. Condensers have different entering air temperatures.

3.11. If the subcooling is not known, charge the system with the correct quantity of refrigerant, then measure and record the resulting subcooling.

3.12. If the system only contains vapor, recover the remaining refrigerant vapor, pressurize the system with nitrogen for a leak test, evacuate the system, and weigh in the refrigerant charge. When the system contains liquid refrigerant, charge refrigerant into the system until the maximum temperature drop is obtained through the evaporator in the a/c unit or heat pump in the cooling mode.

3.13. The safe starting charge is 1.5 lb of refrigerant for each horsepower of the compressor.

CHAPTER 4

4.1.

1. Adjust the refrigerant to the correct compressor operating suction and discharge pressures.

2. Adjust the refrigerant charge to the correct superheat according to the manufacturer's instructions.

3. Adjust the refrigerant charge to the correct subcooling if the correct subcooling is known.

4. Adjust the refrigerant for peak performance.

5. Recover the refrigerant in the system, evacuate the system, and weigh in the calculated refrigerant charge.

4.2. If a means of accurately measuring the cfm flowing through the evaporator is available, the net cooling capacity method is used. If that is not available, determine the gross cooling capacity and subtract the motor heat.

4.3. The formula is cfm divided by Vs (ft^3/lb) multiplied by 60.

4.4. False. Only sensible heat is involved.

4.5. The minimum number of discharge temperature readings required is four.

4.6. False. The center hub of the fan does not move very much air.

4.7. The formula for finding the heat removed or put into the air is Btuh = (cfm)(TD)(1.08).

4.8. The formula for finding motor heat input is Btuh = (volts)(amps)(power factor)(3.413 Btu/watt).

4.9. Power factor may be defined as the true power in watts divided by the apparent power in volt-amperes.

CHAPTER 5

5.1. A comfort zone is the range of temperature and humidity at which the greatest number of people are comfortable.

5.2. People are most comfortable at 80°F dry bulb and 66.5°F wet bulb, or 50% relative humidity.

5.3. Latent heat has a greater effect on the drop in temperature through the evaporator.

5.4. The desired temperature drop would be 17.5°F.

5.5. False. Reduction in cfm increases the drop in temperature.

5.6. False. The evaporator is not 100% efficient.

5.7. False. The cfm is decreased to increase the temperature drop.

5.8. The height should be the venturi opening diameter plus 2 in. The width should be outside diameter of the front of the venturi opening.

CHAPTER 6

6.1. The problems in an air conditioning system are either due to refrigerant quantity or refrigerant flow rate.

6.2. Possible causes could be an overcharge of refrigerant, a flow restriction in the liquid line, or false pressures exist in the system.

6.3. False. The TXV controls the superheat.

6.4. When the setting of a TXV is changed, the only change that occurs is in the operating superheat.

6.5. An elevation in the liquid line reduces the pressure and boiling point of the liquid due to the weight of the liquid.

6.6. The maximum recommended height of the evaporator above the condensing unit is 50 ft.

6.7. Insulation on the liquid line is required when it is in an area in which the temperature is higher than the temperature of the outdoor air.

6.8. If the metering device is a TXV, the valve will close down and compensate for the higher inlet pressure. TXVs also create an oil-return problem.

6.9. If the metering devices are capillary tubes or restrictors, the additional pressure on the refrigerant due to the weight of the refrigerant above the condensing pressure will cause flooding of refrigerant in the evaporator. If the refrigerant charge is reduced to reduce the flooding, the system will have poor subcooling, excessive flash gas, and low system capacity.

6.10. The first trap off the evaporator has an equivalent length of 9 ft.

6.11. The equivalent length of the trap in the suction (vapor) line riser is 12 ft.

Pro-Series
The HVAC/R Professional's Field Guides

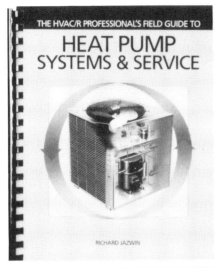

Heat Pump Systems & Service contains service manuals from:
- Bryant/Day & Night/Payne
- Carrier
- Coleman
- Comfortmaker
- Goettl
- Heil-Quaker
- Lennox
- Rheem
- Sun Dial
- Weathertron
- York

You're at the jobsite, and it's been a long, hard fix so far. You think you've all but wrapped it up when you discover that you don't have the right information, and you don't know what to do next. What do you do now? You go to your truck and pull out one of your Pro-Series HVAC/R Professional Field Guides. Problem solved.

Thousands of service technicians all over the world know the importance of having a reliable reference source at the jobsite or in the shop — a reference source like the Pro-Series field guides. These valuable books not only show the ins and outs of ice machines, heat pumps, and gas furnaces, they also contain actual manufacturers' service manuals to give you on-site information about the model you're working on.

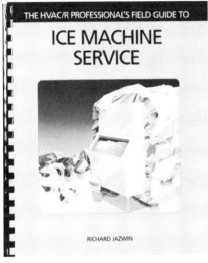

Ice Machine Service contains service manuals from:
- Crystal Tips
- Hoshizaki
- Ice-O-Matic
- Kold-Draft
- Manitowoc
- Remcor
- Ross-Temp
- Scotsman
- SerVend

Technical Books
The books of choice.

Order today!
1-800-837-1037

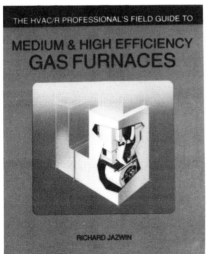

Medium & High Efficiency Gas Furnaces contains service manuals from:
- American Standard
- Arcoaire/Comfortmaker
- Armstrong
- Bard
- Bryant/Day & Night/Payne
- Carrier
- Coleman
- Comfort-Aire
- Heil
- Lennox
- Rheem
- Trane
- York

The ultimate hvac/r reference set!
Reference Notebook Set

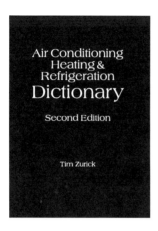 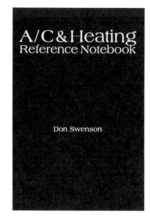 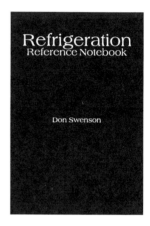

It's new! It's improved! The all-time BNP classic *Reference Notebook Set* has been updated to meet the needs of today's service technician.

With the *Reference Notebook Set*, you can quickly locate important definitions, capillary tube length conversion charts, cooling load factors, pipe sizing tables, and much more.

What makes this set so good is that each book can fit into your pocket, giving you instant information that makes your job easier and more efficient. The set consists of:

• ***Air Conditioning, Heating, and Refrigeration Dictionary, Second Edition*** - An updated, alphabetical listing of all relevant hvac/r terms and their definitions.

• ***A/C & Heating Reference Notebook*** - Brand new and up-to-date numerical information on heating and cooling loads, humidification, ducts and blowers, piping and tubing, as well as measurement conversions, all found in table form for quick reference.

• ***Refrigeration Reference Notebook*** - Completely revamped with updated information on heat load factors, refrigerant properties, pipe and tubing sizing, wire and metal sizing, blower data, conversions, and more!

Whether you are new to the hvac/r fields or have years of experience, the *Reference Notebook Set* is a must have.

Each book can be ordered separately. Call for more information.

Order Today! 1-800-837-1037

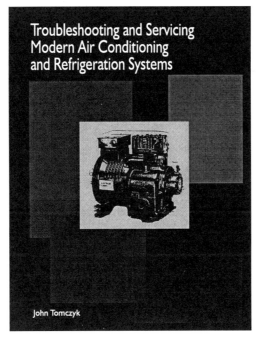

Troubleshooting at its best!

Your work environment is a complex one. Not only must you stay abreast of all new advances in technology, improved service techniques, and environmental regulations, you must have a sound understanding of how many different systems work.

Troubleshooting & Servicing Modern Air Conditioning & Refrigeration Systems, a new book from BNP Technical Books, was written to give service technicians all the information needed to accurately diagnose and solve various system problems. This book will give you a greater understanding of refrigeration and air conditioning while exploring more complex topics and detailed troubleshooting procedures. It is a valuable tool that will give you a helping hand at the jobsite or in the shop.

Inside you'll find the important topics of today — the changes affecting the refrigeration and air conditioning industries with an emphasis on the phaseout of CFC and HCFC refrigerants. You'll get detailed information on the most current leak detection methods, venting regulations, alternative refrigerants, and retrofit guidelines. In addition, you'll get important refrigerant changeover guidelines for the following conversions: R-12 to R-134a; R-12 to MP39; R-502 to SUVA HP80. Hardcover • 282 pages

Troubleshooting & Servicing Modern Air Conditioning & Refrigeration Systems
Order Today! 1-800-837-1037

BNP
Technical Books
The books of choice.

Other Titles Offered by BNP

Troubleshooting and Servicing Modern Air Conditioning and Refrigeration Systems
Fire Protection Systems
Piping Practice for Refrigeration Applications
Born to Build: A Parent's Guide to Academic Alternatives
Refrigeration Fundamentals
Technician's Guide to Certification
Plumbing Technology
Power Technology
Refrigeration Licenses Unlimited, Second Edition
A/C Cutter's Ready Reference
How to Solve Your Refrigeration and A/C Service Problems
Blueprint Reading Made Easy
Starting in Heating and Air Conditioning Service
Legionnaires' Disease: Prevention & Control
Schematic Wiring, A Step-By-Step Guide
Sales Engineering
Hydronics
How to Design Heating-Cooling Comfort Systems
Industrial Refrigeration
Modern Soldering & Brazing Techniques
Inventing from Start to Finish
Indoor Air Quality in the Building Environment
Electronic HVAC Controls Simplified
Heat Pump Systems and Service
Ice Machine Service
Troubleshooting and Servicing A/C Equipment
Stationary Engineering
Medium and High Efficiency Gas Furnaces
Water Treatment Specification Manual, Second Edition
The Four R's: Recovery, Recycling, Reclaiming, Regulation
SI Units for the HVAC/R Professional

TO RECEIVE A FREE CATALOG, CALL

1-800-837-1037

**BUSINESS NEWS
PUBLISHING COMPANY**
Troy, Michigan
USA